Biographical Memoirs of Fellows of the British Academy XVII

Glen Dudbridge

2 July 1938 – 5 February 2017

elected Fellow of the British Academy 1984

by

WILT L. IDEMA

Biographical Memoirs of Fellows of the British Academy, XVII, 1–18
Posted 18 June 2018. © British Academy 2018

GLEN DUDBRIDGE

Born in 1938 in Clevedon, Somerset, Glen Dudbridge attended Bristol Grammar School. Following his National Service, he read Chinese at Cambridge, where he was taught by H. C. Chang, who is probably best known for his substantial and densely annotated anthology of Chinese vernacular literature which appeared in 1973 as *Chinese Literature: Popular Fiction and Drama* (Edinburgh). He also greatly benefited from the expertise of Piet van der Loon, who would remain a major source of inspiration throughout his life. As well as Chang, van der Loon must have drilled him in philology and bibliography. Following his years at Cambridge, Dudbridge continued his studies at the New Asia College in Hong Kong. In 1965, he was appointed as Lecturer in Modern Chinese at Oxford. Twenty years later, in 1985, he was made Professor of Chinese at Cambridge, but returned to Oxford in 1989 when he was appointed in the same function there. He served as Chair of the European Association of Chinese Studies from 1998 to 2002, and was a visiting professor at Yale University, University of California, Berkeley, and the Chinese University of Hong Kong. He was elected a fellow of the British Academy in 1984 and was awarded an Honorary Membership of the Chinese Academy of Social Sciences in 1996. Following retirement from his Oxford Chair, he remained actively involved in academic life and research. He is survived by his wife, their two children and four grandchildren.

The topic of Dudbridge's doctoral dissertation was the development of the legend of the Journey to the West, up to its appearance in 1592 in the form of a 100-chapter novel, nowadays usually ascribed to Wu Cheng'en 吴承恩. His first published articles dealt with the relation of the early editions of this *Xiyou ji* 西游记 to the shorter versions of the legend that circulated at the same time. In a fifty-page article published in 1969, he carefully surveyed the known editions of the full novel and the shorter versions, concluding that the latter were based on the longer version and not the other way around, but that the story of the birth of Xuanzang first made its appearance in the short version edited by Zhu Dingchen 朱鼎臣, and was only later incorporated into the 100-chapter version.[1] That has remained a conclusion difficult to accept for those scholars who grew up reading the novel that included the episode of Xuanzang's birth.[2] Furthermore, Dudbridge's conclusion that the ascription of the 100-chapter novel to Wu Cheng'en is based on the flimsiest of grounds regrettably does not seem to have had much impact.

Dudbridge's first monograph publication was *The* Hsi-yu Chi: *a Study of Antecedents to the Sixteenth-Century Chinese Novel* of 1970.[3] This study fitted in with

[1] G. Dudbridge, 'The Hundred-Chapter *Hsi-yu Chi* and its early versions', *Asia Major*, n.s. 14 (1969), 141–91.
[2] Xu Haoran 许浩然, 'Yingguo hanxuejia Du Deqiao yu *Xiyou ji* yanjiu' 英国汉学家杜德桥与西游记研究, *Zhongnan daxue xuebao*, 18/1 (February 2012), 187–91.
[3] G. Dudbridge, *The* Hsi-yu chi: *a Study of the Antecedents of the Sixteenth-Century Novel* (Cambridge,

the new developments of Western sinology after the Second World War. Academic sinology of the first half of the twentieth century had been focused very much on the philological study of ancient Chinese culture. But the impact of the May Fourth Culture, the influx of young Chinese scholars and the organisation of Chinese studies in the USA in departments of Chinese/East Asian languages and literatures now resulted in a growing interest (first in the USA but later also in Europe) in the vernacular fiction of the last dynasties. Dudbridge's monograph immediately established him as a master in this field. The book basically consists of two parts. In the first, Dudbridge meticulously surveys each known reference to the legend of the Journey to the West and its main characters from the twelfth to the sixteenth century, carefully distinguishing direct quotations from summaries and other indirect references. In the second part of his monograph, he evaluates the different theories that had been put forward to explain the origin of the character of Monkey. Taking as his starting point the image of the Acolyte Monkey (*houxingzhe* 猴行者) of the *Da Tang Sanzang qujing shihua* 大唐三藏取经诗话 (the story, interspersed with poems of how Sanzang of the Great Tang fetched the sutras), the earliest known account of the legend, Dudbridge rejected each and every theory proposed by his eminent predecessors in the field, as these tended to start from the image of Monkey as found in the 100-chapter novel of some centuries later. No wonder that C. T. Hsia, in his review of Dudbridge's work in the *Journal of Asian Studies*, repeatedly mentioned his 'caution' and his 'skepticism'.[4] Hsia also pointed out the discrepancy between Dudbridge's opening chapter and the main content of the book. In this opening, Dudbridge dwells at some length on the Parry-Lord theory of oral composition, which in the 1960s and 1970s was at the height of its popularity, stressing that written references to the legend were only an infinitesimal fraction of the rich and constantly changing but also unknowable legend as it was orally transmitted. While Hsia could at times be quite blunt—if not to say abusive—in his comments on the works of other scholars in the field of vernacular literature, here he expressed himself very mildly, but still Dudbridge took issue with this review.[5]

In an article of 1988, 'The *Hsi-yu chi* Monkey and the fruits of the last ten years',[6] Dudbridge would return to the issue of the origin of Monkey and evaluate the theories that had been put forward in the years since he published his monograph. While

---

1970): reviewed by C. T. Hsia in *Journal of Asian Studies*, 30 (1971), 887–8; and A. C. Yu in *History of Religions*, 12 (1972), 90–4.

[4] In *Journal of Asian Studies*, 30 (1971), 887–8.

[5] G. Dudbridge, 'The *Hsi-yu chi* Monkey and the fruits of the last ten years', *Journal of Asian Studies*, 31 (1972), 351.

[6] Originally published in *Chinese Studies*, 6 (1988), 463–86; reprinted in G. Dudbridge, *Books, Tales and Vernacular Culture: Selected Papers on China* (Leiden, 2005), pp. 254–74.

still displaying the same sceptical attitude to the various proposals advanced, mostly by Japanese scholars, he also tries to find an explanation for the function of Monkey by looking at the role of this figure in funerary rituals of the late nineteenth century in Fujian. Here Monkey acted as the guardian of the soul during its transition to the other realm.

Dudbridge's second monograph appeared in 1978 as *The Legend of Miaoshan*.[7] The research must have been concluded some years earlier, as Victor Mair pointed out in his review that the bibliography contained no publications later than 1973. This of course meant that Dudbridge had conducted his research during the heyday of the Great Proletarian Cultural Revolution in China. In view of the major role of the bodhisattva Guanyin 观音 in the Journey to the West, the choice of the legend of her female incarnation as the princess Miaoshan 妙善 was perhaps an obvious one for a follow-up project. For Dudbridge, one of the major attractions of this legend was that here it was possible, in contrast to the frustrating Journey to the West, to pinpoint the moment at which this legend entered Chinese culture in a fully developed shape. Dudbridge, otherwise writing quite concisely, describes at some length how the metropolitan official Jiang Zhiqi 蒋之奇 (1031–1104), demoted to the provincial posting of prefect of Ruzhou in Henan, in early 1100 visited the Xiangshan Monastery in Baofeng—by that time already a well-established centre for Guanyin veneration—and there was presented by the abbot Huaizhou 怀昼 with a biography of Miaoshan, which had apparently been brought to the monastery just a few days earlier by a mysterious monk who had since departed without leaving a trace. The biography, said to have been retrieved from a pile of waste paper, claimed to be the tale of the bodhisattva's incarnation as told to the saintly seventh-century monk Daoxuan 道宣 (596–667) by an unspecified heavenly being and recorded by one of Daoxuan's disciples. Jiang Zhiqi copied out the text and prefaced it with an account of his meeting with the abbot; and a stele of this text, written out in the calligraphy of Cai Jing 蔡京 (d. 1126), was erected in the temple later that same year, to be reinscribed in 1304. Needless to say, Dudbridge received no reply from the local authorities in China when he wrote to them for information on the whereabouts of the stone and requested a rubbing in case it was still available.

As he had no access to the stele in the early 1970s, Dudbridge reconstructed its contents on the basis of two later summaries. Its main difference with later versions of the legend was that this early version did not as yet include an account of Miaoshan's

---

[7] G. Dudbridge, *The Legend of Miaoshan* (London, 1978): reviewed by W. L. Idema in *T'oung Pao*, 66 (1980), 286–8; by V. H. Mair in *Harvard Journal of Asiatic Studies*, 39 (1979), 215–18; by A. Seidel in *Journal of Asian Studies*, 38 (1979), 770–1; and by K. Whitaker in *Bulletin of the School of Oriental and African Studies*, 42 (1979), 193–4.

visit to the underworld and her liberation of all sinners. This episode was first included in a version that was connected to the name of Guan Daosheng 管道昇 (1262–1319), the wife of the famous calligrapher Zhao Mengfu 赵孟頫 and a fine painter and calligrapher in her own right. With the addition of this episode the legend had, in Dudbridge's view, acquired its mature form. He continued his survey of the historical development of the legend with a discussion of the *Xiangshan baojuan* 香山宝卷 (Precious Scroll of Incense Mountain, which carries a preface claiming to date from 1103), the late sixteenth-century novel *Nanhai ji* 南海记 (The Story of the Southern Sea, composed by Zhu Dingchen, most likely on the basis of the precious scroll) and a seventeenth-century rewriting of the legend, again in the precious scroll format (first studied by the Dutch sinologist Henri Borel), but did not pursue the many adaptations of the legend in local forms of drama and storytelling. In the final two chapters, he preferred to look into some of the background materials of the legend (especially the Lotus Sutra and folklore, comparing the legend of Miaoshan to *King Lear*), and into its ritual functions (filial piety and salvation).

Dudbridge researched his *The Legend of Miaoshan* when the large databases he would happily use in his later work were still in the future. When his book came out and was reviewed in Taiwan, Lai Ruihe 賴瑞和 pointed out that not only was the short text of Guan Daosheng's life of Miaoshan available, but also that a rubbing of the second half of Jiang Zhiqi's text as erected at the Upper Tianzhu Monastery near Hangzhou in 1104 had been preserved.[8] Dudbridge then quickly published an article in the *Harvard Journal of Asiatic Studies* in which he evaluated the importance of these new materials to his work,[9] and when in 1990 a Chinese translation of his study appeared in Taiwan these findings were incorporated into the text.[10] In this way, his monograph also became known in the People's Republic of China, but only circulated in very small numbers. Writing in 2011, Chen Yongchao 陈泳超 of Peking University states that he had to make special efforts to get hold of the book.[11] In his long and judicious review of Dudbridge's monograph, he puts his work alongside Gu Jiegang's

---

[8] Lai Ruihe, 'Miaoshan chuanshuo de liangzhong xin ziliao' 妙善傳說的兩種新資料, *Zhongwai wenxue*, 9 (July 1982), 116–26.

[9] G. Dudbridge, 'Miaoshan on stone: two early inscriptions', *Harvard Journal of Asiatic Studies*, 42 (1982), 589–614.

[10] Du Deqiao 杜德橋, *Miaoshan chuanshuo: Guanyin pusa yuanqi kao* 妙善傳說觀音菩薩起考, trans. Li Wenbin 李文彬 (Taipei, 1990). The translation was accompanied by a full photographic reprint of a rare, early seventeenth-century printing of the *Nanhai ji* from the Bodleian Library.

[11] Chen Yongchao, '"Xieben" yu chuanshuo yanjiu fanshi de bianhuan: Du Deqiao *Miaoshan chuanshuo* shuping' 写本与传说研究范式的变换杜德桥妙善传说述评, *Minzu wenxue yanjiu* (2011.5): 5–17. For an earlier evaluation of Dudbridge's monograph see Dong Xiaoping 董晓萍 'Chuanshuo yanjiu de xiandai fangfa yu xiandai de wenti: ping Du Deqiao de *Miaoshan chuanshuo*' 传说研究的现代方法与现代的问题：评杜德桥的妙善传说, *Minzu wenxue yanjiu* (2003.3), 3–13.

顾颉刚 (1893–1980) study on the legend of Meng Jiangnü 孟姜女 in importance. Chen would appear not to have been aware that in 2004 Dudbridge had published a revised edition of his *The Legend of Miaoshan*, which not only incorporated the conclusions of his 1992 article, but also could make use of the full text of the (slightly damaged) text by Jiang Zhiqi as reinscribed in 1304. While Dudbridge was not allowed to see the stone at its current location during a visit to Baofeng in the late 1990s, he now had access to a rubbing and photographs provided by Chinese colleagues. His letter for information on the stele in the early 1970s apparently had been received after all, because local Chinese sources now reported that the stele had been exhibited in Oxford in that period (and would continue to do so despite Dudbridge's protests that such had not happened at all).

It is very unlikely that the text that Huaizhou showed to Jiang Zhiqi did indeed date from the Tang dynasty, let alone that it had been revealed to Daoxuan who was well known to have visions. Dudbridge clearly implies that the life of Miaoshan may have been concocted by the abbot to bamboozle an eager devotee of Guanyin. But recently, voices have been heard that argue that the text after all may well have been authentically revealed, even if not from the Tang. Whatever the truth of the matter, Dudbridge would spend most of his research of the following decades on authentic texts of the seventh to tenth centuries, especially those that had been preserved in the huge compendia compiled during the reign of Taizong of the Song (r. 976–997), such as the *Taiping yulan* 太平御览 (Imperial Reader for the Time of Supreme Peace) and the *Taiping guangji* 太平广记 (Extensive Records for the Time of Supreme Peace). In view of his meticulous attention to each detail of the texts he was working with, Dudbridge would not only deal with issues of authorship and date of composition but also with all aspects of their transmission. This meant that he was not only interested in the transmission of these texts up to the moment they were included in the Song imperial collections, but also in the printing history of the *Taiping guangji*. And because Dudbridge soon was convinced that the *Taiping guangji* had been hastily and shoddily compiled, he also was fascinated by the occasional transmission of classical tales of the Tang outside that collection, whether in full or synoptic versions.

Several collections of translations of classical tales from the Tang had appeared by the 1980s, but Dudbridge set a new and much higher standard for the study of these materials with the publication of his *The Tale of Li Wa: Study and Critical Edition of a Chinese Story from the Ninth Century* of 1983.[12] Like his *The Legend of Miaoshan*,

---

[12] G. Dudbridge, *The Tale of Li Wa: Study and Critical Edition of a Chinese Story from the Ninth Century* (London, 1983): reviewed by W. L. Idema in *T'oung Pao*, 71 (1985), 279–82; by W. H. Nienhauser in *Journal of the American Oriental Society*, 106 (1986), 400–2; by D. E. Pollard in *Journal of the Royal Asiatic Society*, 116 (1984), 304–5; and by G. Weys in *Bulletin of the School of Oriental and African Studies*, 48 (1985), 172–3.

this volume was published in the 'Oxford Oriental Monographs' series by Ithaca Press, and it is difficult to imagine that he could have published this work in this form with any other publisher, as he starts his 'Introduction', without any preamble, with a highly technical discussion of the textual history of the tale. This monograph definitely was not written to win over undergraduates, but to teach his fellow scholars a lesson. That lesson was that tales such as *Li Wa zhuan* might have been written as light entertainment but were composed by young men who had passed the *jinshi* examination and for their friends who had the same background, and that we might therefore expect them to draw on their shared readings for their language, whether as subtle allusions or trite clichés. Very well aware that he might indulge in over-annotation, Dudbridge provides a near exhaustive list of all borrowings from the classics and the *Wenxuan* 文选, among which especially the borrowings from the *Zuozhuan* 左传 stood out. But Dudbridge does not limit himself to providing a new critical edition of the Chinese text of this tale and a new translation; he also highlights the problematical nature of its traditional ascription to Bai Juyi's 白居易 (772–846) younger brother Bai Xingjian 白行简 (776–826), relates the tale to the sensational history of the Zheng 郑 family in the wake of the An Lushan rebellion, discusses the background and status of Li Wa 李娃 as a private courtesan and the possibility of her marriage to her paramour, and traces the later adaptations of the tale in anecdote, *huaben* and drama.[13] Those who might think that Dudbridge had lost his interest in religion by turning to this story of love and betrayal, family disruption and reunion, would find that this was not the case at all, as here (as elsewhere) he relies heavily on the ritual theory of the French anthropologist and folklore scholar Arnold van Gennep (1873–1957) on 'rites of passage' in analysing the literary structure of the tale.

This abiding interest in religion in literature is also evident in some other fine readings of the individual tales that Dudbridge published in his period. In his wonderful reading of the *Liu Yi zhuan* 柳毅传 and its analogues, for instance, he introduces the theme of the ghost marriage.[14] This focus on religion was also quite apparent in a number of articles that were written as side products of his work on the reconstruction and interpretation of Dai Fu's 戴孚 *Guangyi ji* 广异记 (Great Book of Marvels), a collection from the third quarter of the eighth century, a project that would eventually result in his *Religious Experience and Lay Society in T'ang China* of 1995.[15] Dudbridge

---

[13] Dudbridge goes on to suggest that the popularity of the self-sacrificing courtesan in traditional Chinese literature might have facilitated the Chinese reception of *La Dame aux Camelias* in its translation by Lin Shu in the early years of the twentieth century.

[14] G. Dudbridge, 'The tale of Liu Yi and its analogues', in E. Hung (ed.), *Paradoxes of Traditional Chinese Literature* (Hong Kong, 1994), pp. 61–88; reprinted in Dudbridge, *Books, Tales and Vernacular Culture*, pp. 151–79.

[15] G. Dudbridge, *Religious Experience and Lay Society in T'ang China: a Reading of Tai Fu's* Kuang-I chi

had already included a translation of the preface to this collection by Dai Fu's friend Gu Kuang 顾况 (d. 806) in the 'Introduction' of his *The Tale of Li Wa*. Dai Fu had been a minor official, mostly active in the Zhejiang area. Dudbridge extracted over three hundred items from his collection from the *Taiping guangji*, which showed Dai Fu to be a precise observer of exceptional events in his surroundings, as well as an eager recorder of strange stories that were told to him by friends and acquaintances.[16] In these ways, Dudbridge argues, Dai Fu's tales were not works of 'creative fiction' but preserved 'the oral history of a remote age'. Dudbridge studied the many voices recorded in the *Guanyi ji* 'not really to build up a knowledge of events and institutions with documentary data, but rather to explore the perceptions of that long dead generation as it confronted the visible and invisible world all around'.[17]

To analyse these rich materials, Dudbridge coins the terms 'inner story' and 'outer story'. The inner story refers in his usage to a personal supernatural experience such as a dream, a vision or a revelation, and also a legend, which will be culturally conditioned but is basically beyond verification; whereas the outer story concerns the equally culturally conditioned ways in which society at large publicly deals with these exceptional experiences of one or more of its members.[18] In contrast to the professional religious literature as collected in the Tripitaka or the Daozang, which is made up of writings by clerics for clerics, Dudbridge argues, these tales provide us with a quite reliable record of how Chinese society of the third quarter of the eighth century dealt with the irruption of the divine in their daily lives.[19] After demonstrating the usefulness of his distinction between inner story and outer story in his first chapter, Dudbridge proceeds with a detailed analysis of Gu Kuang's preface to the *Guangyi ji*

(Cambridge, 1995): reviewed by R. F. Campany in *Chinese Literature: Essays, Articles, Reviews*, 19 (1997), 143; by Huang Chi-liang in *China Review International*, 5 (1998), 120–4; by R. Kirkland in *Journal of Asian Studies*, 55 (1996), 977–8; by O. Moore in *Journal of the Royal Asiatic Society*, 7 (1997), 494–5; and by W. H. Nienhauser in *T'oung Pao*, 85 (1999), 181–9. For Chinese evaluations of this monograph see Xu Haoran, 'Yingguo Hanxuejia Du Deqiao dui *Guangyi ji* de yanjiu' 英国汉学家杜德桥对广异记的研究, *Shixue yuekan* (2011.07), 134–6; and Yang Weigang 杨为刚, 'Zhiguai xiaoshuo yanjiu de yuwai zhi yan: Du Deqiao *Zongjiao tiyan yu Tangdai shisu shehui: Guangyi ji* de yizhong jiedu pingdu' 志怪小说研究的域外之眼：杜德桥宗教体验与唐代世俗社会广异记的一种解读评述, *Huawen wenxue*, 110 (2012.3): 31–4. Both articles place Dudbridge's study in the context of the history of mentalities.

[16] Summaries of all items are provided in an appendix. For the reader's convenience, these items are numbered according to the order of presentation in the edition of the text by Fang Shiming 方詩銘 in *Mingbao ji, Guangyi ji* 冥暴記廣異記 (Beijing, 1992).

[17] Dudbridge, *Religious Experience and Lay Society in T'ang China*, p. 6. Dudbridge points out that J. J. M. de Groot frequently made use of materials from *Guanyi ji* in his *The Religious System of China*.

[18] Ibid., pp. 14–15. Nienhauser notes in his review that Dudbridge is not always able to maintain this distinction between 'inner' and 'outer'.

[19] For all his work on popular religion of the Tang dynasty, Dudbridge showed little or no interest in the *bianwen* literature of Dunhuang, in which the clergy preach to the laity.

in his second chapter. Following a chapter on the life and times of Dai Fu, Dudbridge continues his analysis of the materials provided by that writer through a number of thematic chapters, such as 'The Worshippers of Mount Hua' which considers the unavoidable encounters of individual men and women with the amorous male and female deities of this holy mountain.

Dudbridge's wide reading in classical tales inspired him with a mission to reconstruct the lost works that could be (at least partially) salvaged from the early Song compendia and in this way reconstruct the lost individual voices of authors whose works had not been transmitted independently. He describes this mission in his 1999 Panizzi Lectures at the British Library, *The Lost Books of Medieval China*. In the first of the three talks collected in this slim volume, he sets out his conclusions regarding the sources, the compilation and the resulting quality of the *Taiping yulan* and the *Taiping guangji*. At the same time, he also discusses the value of the Song dynasty bibliographies in asserting the nature and organisation of texts that no longer exist as independent works but were in some cases still available in such form to the compilers of these catalogues. The second and third talks do not return to his work on Dai Fu and his *Guangyi ji*, but deal with two other cases. The first of these is the *Sanguo dian-lüe* 三国典略 (Summary documents of three kingdoms), a chronological history of the sixth century up to the foundation of the Sui dynasty compiled in the eighth century by a man named Qiu Yue 丘悦. Dudbridge had compiled a critical edition of the surviving items of this text (in this case also strongly relying on Sima Guang's 司马光 *Zizhi tongjian* 资治通鉴) in co-operation with the Chinese scholar Zhao Chao 赵超 of the Academy of Social Sciences, an edition that was published in 1998 in Taipei.[20] In describing the aim of this project, Dudbridge says in his Panizzi Lectures:

> The point has already been made that a project like this aims to do more than just recovering bits of text. Among other things it will look for serious new insights into Chinese history. From the T'ang period and before, outside the standard histories, we have very few surviving historical records. To recover something like this from the hand of a private historian should give a rare chance to get behind the bureaucratically inspired values and choices of the imperial historian, and discover other values, other choices, and different information.[21]

To bring out the distinctive approach of Qiu Yue, Dudbridge covers two episodes in some detail. The first concerns the transferal of the Liang capital from Jianye (Nanjing) to Jiangling, in which the last Liang emperor allowed himself to be swayed by the voice of a soothsayer against the advice of all his officials. The second deals with the

---

[20] Qiu Yue, *Sanguo dianlüe jijiao* 三國典略輯校, ed. Du Deqiao (G. Dudbridge) and Zhao Cao (Taipei, 1998).

[21] G. Dudbridge, *Lost Books of Medieval China* (London, 2000), pp. 31–2.

burning of the imperial library at the end of that dynasty, whether by accident or design.

The third and final talk of the Panizzi Lectures is devoted to the *Liang sigong ji* 梁四公记 (Four Gentlemen of the Liang), a short text of the early eighth century and uncertain authorship. It describes the adventures of four mysterious men at the court of Emperor Wu of the Liang. They arrive in the capital in rags, yet impress not only the court officials but also the emperor by their superior abilities in every branch of knowledge. Three fragments of this text have been included in the *Taiping guangji*, of which the first is quite extensive and would appear to provide a complete narrative, leaving Dudbridge wondering where the two other fragments should fit in. Dudbridge also expresses his bewilderment at the original intention of this work, whether we should read it as a satire (perhaps of the court of Xuanzong) or as 'a work of exuberant fantasy', and in that context once again expresses his criticism of modern Chinese scholarship on Tang classical tales that—caught in the binary contrast of *zhiguai* 志怪 (anomaly records) and *chuanqi* 传奇 (stories of the strange)—could discuss a work such as the *Liang sigong ji* only as a precursor of later full-fledged *chuanqi*.

This criticism was only one of the many manifestations of Dudbridge's unease with these terms. His extensive and long-lasting engagement with the classical tales of the Tang had set him off on a crusade against *chuanqi*, that is to say, against the preferential treatment by literary historians of a small group of tales that were seen (following Lu Xun 鲁迅) as the culmination of the development of the classical tale towards self-conscious fiction. In many talks and publications, Dudbridge argued that this procedure not only resulted in a reductive reading of these few texts that had been classified as *chuanqi*, but that it also resulted in wilful neglect of the overwhelming majority of tales that were put away as *zhiguai* or *yishi* 轶事 ('apocryphal anecdotes'). Little of this unease over the use of *zhiguai* and *chuanqi* was as yet discernable in his monograph on *Li Wa zhuan*, but in the following years Dudbridge became increasingly outspoken on this issue, stressing that there were no hard and fast criteria to distinguish *chuanqi* from the great mass of classical tales and anecdotes, and that many of the usually ignored tales not only had considerable literary merits but also many other characteristics that made them worth studying. Dudbridge provides a clear statement of his concerns in the introductory paragraphs of his 'A question of classification in Tang narrative: the story of Ding Yue', an article that was first published in 1999. Having pointed out the May Fourth mistake of equating the traditional Chinese notion of *xiaoshuo* ('small talk') with fiction and seeking for 'onward progress' in literature, he writes:

> For Tang narrative one important category was bequeathed by Lu Xun: since his time the label *chuanqi* has clung stubbornly to his anthology pieces from the Tang and Song. The name had no generic status or function in the periods when those stories

were written—a proposition which no one actually denies. Yet even today writers on Chinese narrative are content to go on applying the term mechanically to that small corpus of stories . . . . Stories outside the corpus are often categorized as *zhiguai* (another ancient term resurrected for purposes of modern classification), and as *yishi*, 'apocryphal anecdotes', and likewise subdivided by subject matter.

    By pointing out this pattern of categories within a general habit of classificatory thinking, mutable and unstable as they are, the present paper aims to stress how little the whole system illuminates the literature itself. As students of China we should reach beyond the inherited categories of one or another generation of anthologists or literary historians: we should confront the primary texts as best we can in their own environment and accept all the complexity that may face us there.[22]

For Dudbridge, this is very strong and passionate language indeed. The only problem is of course that many of his colleagues in literature were reading the classical tales of the Tang with thematic interests that were different from his own, such as the emergence of romance. At the same time, the May Fourth Movement that had raised the status of traditional narrative had also taught Chinese intellectuals to despise popular religion as superstition—and the anthropologists who worked on traditional religion were only rarely interested in the history of that tradition. Moreover, while it would appear that the young authors of a flurry of recent monographs on Tang tales and anecdotes have heeded Dudbridge's urgent appeal to draw on as wide a selection of materials as possible, they do so from a non-religious perspective, and only rarely follow his example in reconstituting individual collections and studying their individual characteristics.

    For yet another magisterial demonstration of what may be achieved by doing so, however, we only have to turn to Dudbridge's last monograph, *A Portrait of Five Dynasties China: from the Memoirs of Wang Renyu (880–956)*, which came out in 2013.[23] Wang Renyu 王仁裕 hailed from Qinzhou and served under the Former Shu dynasty; when that regime was overthrown, he served the Later Tang and its successor states until his death. A prolific author, his collected works counted 685 scrolls upon his death, but hardly anything of his formal writings survives. Wang Renyu's anecdotal writings enjoyed a better fate. The *Taiping guangji* includes over two hundred items of Wang's collections of tales and anecdotes, *Yutang xianhua* 玉堂闲话 (Tabletalk from the Hanlin Academy) and *Wangshi jianwen lu* 王氏见闻录 (Things Seen and Heard by Mr Wang):[24] 'Both books offer testimony and comment in Wang Renyu's

---

[22] G. Dudbridge, 'A question of the classification in Tang narrative: the story of Ding Yue', in A. Cadonna (ed.), *India, Tibet, China: Genesis and Aspects of Traditional Narrative* (Florence, 1999), pp. 157–8. The article was reprinted in Dudbridge, *Books, Tales and Vernacular Culture*, pp. 192–213.
[23] G. Dudbridge, *A Portrait of Five Dynasties China: from the Memoirs of Wang Renyu (880–956)* (Oxford, 2013).
[24] Summaries of all these items are provided in an appendix.

own voice, and both equally offer pen-portraits of individuals, gossipy anecdotes, historical memories, legends attached to particular places, and the type of stories we would now call urban myths.'[25] Translating a wide selection from these with detailed annotations, Dudbridge is able to present the 'memoirs' of someone who personally lived through one of the most chaotic and violent periods of Chinese history and was eyewitness to some of the most traumatic events. After the scene has been set in the first chapter, the second chapter deals with the oral traditions that developed around some of these happenings of the late ninth and early tenth century. While religion cannot be expected to play the same role in this volume as it did in the monograph on Dai Fu, the third chapter discusses 'A World of Signs and Symbols'. Chapters four and five deal with Wang Renyu's experiences in Shu and the people he encountered there, whereas chapter six is mostly taken up with the full translation of a long description of the fall of the Shu regime. The remaining chapters deal with Wang's life at the central courts. Chapter eight is devoted to anecdotes concerning 'Music and Musicians' and chapter nine is titled 'The Wild', concerning hunting stories and animals. Before them, chapter seven is entitled 'The Khitan'—Dudbridge writes:

> The Khitan-related memoirs come down to us like nearly all the rest through the *Taiping guangji*. But three of them, the most important of the group, share an unusual circumstance. They vanished from the Chinese transmission of *Taiping guangji* and survive for us to read only in the Korean text *T'ae p'yŏng Kwang ki sang chŏl*.

This last title refers to a (partially preserved) Korean selection from the *Taiping guangji* that was printed in 1467 (so one hundred years earlier than the first preserved Chinese edition of *Taiping guangji*). Dudbridge continues:

> What might that signify? We have seen that the early transmission of *Taiping guangji* in China before 1567 lies in shadow. But concealed in that shadow is the bulky presence of the Mongol Yuan dynasty, whose direct control covered China but not the Korean peninsula. It is irresistibly tempting to guess that the three 'Khitan' memoirs, filled with vigorous anti-barbarian sentiments, were deemed unwelcome and dispensable during the Mongol era, yet escaped the same attention in Korea, where an earlier edition was probably handed down. In any case, we are lucky to have them.[26]

If Dudbridge is right in his suggestion that the *Taiping guangji* was censored in China during the Yuan, it becomes of course an even more problematical source than it already is on account of its hasty and shoddy compilation.

Even though Dudbridge's research since the 1980s was focused on the classical tale of the Tang and Five Dynasties period, he remained at the same time very much inter-

---

[25] Dudbridge, *A Portrait of Five Dynasties China*, p. 5.
[26] Ibid., p. 146.

ested in Chinese vernacular and popular literature (and its relation to popular religion) of the late imperial period. We have already mentioned his 'The *Hsi-yu chi* Monkey and the fruits of the last ten years' of 1988, and his ongoing work on the legend of Miaoshan. His participation in the 1989 conference on 'Pilgrims and Sacred Sites in China' resulted not only in the translation of chapters 68 and 69 of the seventeenth-century novel *Xingshi yinyuan zhuan* that was published in the conference volume as 'Women pilgrims to T'ai Shan: some pages from a seventeenth-century novel',[27] but also in a detailed study of these chapters that was separately published in *T'oung Pao* as 'A pilgrimage in seventeenth-century fiction: T'ai-shan and the *Hsing-shih yin-yüan chuan*'.[28] His work on the amorous deities of Mount Hua of the Tang also stimulated him to pursue the development of the legend of Chenxiang 沉香, especially in the Cantonese ballads known as *muyushu* 木鱼书, in a very detailed article.[29] He also edited a number of late nineteenth-century articles on the aboriginal population of Taiwan.[30]

Dudbridge used the opportunity of the opening of the Institute for Chinese Studies at the University of Oxford on 1 June 1995 to sketch his vision of the future of Chinese studies in a lecture entitled 'China's Vernacular Cultures'.[31] While acknowledging the inevitability in many cases of a top-down study of Chinese culture on the basis of materials prepared by the political and cultural elite at the centre, Dudbridge made a plea for the equal role of the study of regional cultural traditions on the basis of local materials in order to do justice to the richness and variety of Chinese culture in all its complexity, past and present. In 2002, Dudbridge and Frank Pieke also initiated the series *China Studies* with the Leiden publisher Brill. The nigh on forty volumes that have appeared in this series so far have covered a wide range of topics, from traditional fiction to migrant communities in present-day Beijing.

With his mastery of both modern and classical Chinese, his formidable scholarship and his demanding standards, Dudbridge could be a forbidding teacher. But the stern

---

[27] G. Dudbridge, 'Women pilgrims to T'ai Shan: some pages from a seventeenth-century novel', in S. Naquin and Chün-fang Yü (eds.) *Pilgrims and Sacred Sites in China* (Berkeley, CA, 1992), pp. 39–64.

[28] G. Dudbridge, 'A pilgrimage in seventeenth-century fiction: T'ai-shan and the *Hsing-shih yin-yüan chuan*', *T'oung Pao*, 77 (1991), 226–52. Reprinted in Dudbridge, *Books, Tales and Vernacular Culture*, pp. 275–302, as 'A pilgrimage in seventeenth-century fiction: Taishan and the *Xingshi yinyuan zhuan*'.

[29] G. Dudbridge, 'The goddess Hua-yüeh San-niang and the Cantonese ballad *Ch'en-hsiang T'ai-tzu*', *Chinese Studies/Hanxue yanjiu*, 8 (1990), 627–46. Reprinted in Dudbridge, *Books, Tales and Vernacular Culture*, pp. 303–20.

[30] G. Taylor, *Aborigines of South Taiwan in the 1880s: Papers by the South Cape Lightkeeper*, ed. Glen Dudbridge (Taipei, 1999).

[31] G. Dudbridge, *China's Vernacular Cultures: an Inaugural Lecture Delivered before the University of Oxford on 1 June 1995* (Oxford, 1996). Reprinted in Dudbridge, *Books, Tales and Vernacular Culture*, pp. 217–37.

appearance hid a genial and friendly personality. His DPhil students remember their days with him most fondly.[32] I personally could observe his efforts on behalf of international students in the 1980s and 1990s, when the ERASMUS programme enabled European universities to set up networks for the exchange of students in specific fields. Oxford and Cambridge participated in a network for Chinese studies that was co-ordinated from Leiden. In view of their fine facilities at home, the number of British students that were interested in spending a year on the Continent tended to be small, whereas there were always many continental students eager to spend a year at Oxford or Cambridge. Each year, Dudbridge went to great lengths to ensure that at least one continental student could come to Oxford, and that he or she would be housed in one of the colleges to make sure that they would share in the full Oxford experience. In many other ways he also showed his concern for the well-being of these special students.

As a member of the British Academy, Dudbridge played an active and important role in initiating and developing academic co-operation between China and the United

[32] The Festschrift published on the occasion of Dudbridge's retirement from the Chair at Oxford contained contributions by Alan Barr, Li-ling Hsiao, Chloë Starr, Alison Hardie, Rana Mitter, Carolyn Ford, Mark Strange and Daria Berg: D. Berg (ed.), *Reading China: Fiction, History and the Dynamics of Discourse, Essays in Honour of Professor Glen Dudbridge* (Leiden, 2007).

Kingdom. His service to the British Academy had started in October 1979, even before he was elected a Fellow, on the occasion of the Academy's first delegation to China. The delegation included five Fellows (Alec Cairncross, Raymond Firth, James Joll, Toby Milsom and William Watson) and the Academy's Secretary. Because this group did not include a good Mandarin speaker, Dudbridge was invited to go with them. He was much younger than the rest of the delegates and consequently found himself used rather as if he were a member of staff, sorting out the various issues that arose. But from his point of view it was a tremendous opportunity—this was, after all, quite an early high-level humanities and social sciences academic delegation after the Cultural Revolution and Mao's death in 1976—and the delegation was received in the Great Hall by Deng Xiaoping (Dudbridge is at the far right in the back row in the photo on p. 15). The visit would lead, in 1980, to the signing of an exchange agreement with the Chinese Academy of Social Sciences, one of the first with a Western country (the Academy itself had only been established in 1977, when it was separated from the Chinese Academy of Sciences).

After he was elected a Fellow of the British Academy, Dudbridge sat on the China Selection Panel (responsible for the administration of the China exchange agreements) from 1987 to the end of 1997, serving as chair (after Alec Cairncross) from 1990 onwards. He also served on the Academy's Overseas Policy Committee from 1988 to 1995, where he proved to be a thoughtful and valuable member, willing and able to apply his experience and understanding to issues beyond his own specific interests. He also would go on two further British Academy delegations to China. In 1993, he was on the delegation led by Charles Feinstein, along with John Goldthorpe and Marilyn Strathern. Because Dudbridge was unwilling to fly internally in China, the trip involved long train journeys (Beijing to Xian, Xian to Chengdu) which were fascinating for his sociologist and social anthropologist companions. And in 1997, he was part of the delegation led by Tony Wrigley, along with Barry Supple and Jessica Rawson, which went to Beijing, Shanghai, Hong Kong (for talks with the KC Wong Foundation) and Taipei. On both these trips, Dudbridge was crucial in interpreting not merely the language but also the historical, social, cultural and academic contexts, in such a way as to help those members of the team who were not specifically sinologists, and his expertise was regularly deferred to.[33]

***

When David Pollard, well known for his work on modern Chinese literature and the essay, reviewed Dudbridge's *The Tale of Li Wa*, he wrote:

---

[33] These two paragraphs on Dudbridge's service to the British Academy were contributed by Jane Lyddon, former head of International there.

I had almost forgotten what satisfaction and pleasure could be got from reading a work of good old-fashioned sinology. That satisfaction derives from following, at a remove, the patient assembly and collation of texts and commentaries from the libraries of the world, the methodological checking of the hard evidence upon which arguments have been based, taking nothing for granted in the process, and a cool appraisal of the legitimate limits of inference and speculation. The pleasure lies in being party to an imaginative reconstruction of what is known and enlargement of what is thought, still within the bounds of plausibility.[34]

It is clear from these phrases that for Pollard 'good old-fashioned sinology' referred to the philological scholarship associated with the European tradition of sinology of the first half of the twentieth century. In a way, however, Dudbridge was even more 'old-fashioned' than that, because the China scholar he most often explicitly engaged with, especially in his *The Legend of Miaoshan* and in his *Religious Experience and Lay Society in T'ang China*, was the Dutch sinologist J. J. M. de Groot (1854–1921). For all his fame in his own time, de Groot was hardly mentioned in Leiden in my student days, as my teachers considered him at best a ghost from the past of no relevance whatsoever to their own work.[35] Dudbridge most probably had been introduced to the works of de Groot and their combination of ethnographical field-work with historical background studies by Piet van der Loon, who, like de Groot, was fascinated by the popular religious and literary traditions of south-east China. But Dudbridge was modern and unique in applying his philological skills to materials that were ignored by the sinologists of preceding generations, and stood apart from many of his colleagues by not searching for a 'system' or 'synthesis' but by his fascination with personal voices from the past.[36] At the same time, his interpretation of these voices was always informed by his broad reading in the social sciences and criticism.

Rereading Dudbridge's major publications in preparation for this memoir was both a pleasurable and a humbling experience. Of course I had read his works on first appearance, and consulted them on later occasions to my benefit, but reading them again from cover to cover I was not only impressed anew by his sure command of his

---

[34] D. E. Pollard, 'Review of G. Dudbridge, *The Tale of Li Wa*', *Journal of the Royal Asiatic Society*, 116 (1984), 304–5.

[35] So, they were quite surprised when in the late 1960s and early 1970s the British anthropologist Maurice Freedman (1920–1975) showed considerable interest in the works of de Groot, for instance in his 'On the sociological study of Chinese religion', in A. P. Wolf (ed.), *Religion and Ritual in Chinese Society* (Stanford, CA, 1974), pp. 19–41. Freedman joined Oxford University in 1970.

[36] Dudbridge's work on Tang tales shows many similarities to the work of Robert Ford Campany on the classical tales of the pre-Tang period, as the latter likewise stresses that these tales should be seen as historical sources and not as failed precursors of fiction. Campany, however, has a background in religious studies and does not eschew synthesis.

sources and the careful presentations of his findings, but also became aware of how much I had failed to notice earlier. Avoiding the use of fashionable jargon, Dudbridge's works show no signs of aging. With their unique combination of fine textual scholarship, extensive translations and probing analysis of detail, these publications will, I am sure, continue to inspire future generations of students of Chinese society and culture at large.

Note on the author: Wilt L. Idema is Research Professor of Chinese Literature, Harvard University, and Professor Emeritus of Chinese Language and Literature, Leiden University.

Mary Brenda Hesse

15 October 1924 – 2 October 2016

elected Fellow of the British Academy 1971

by

NICHOLAS JARDINE

Biographical Memoirs of Fellows of the British Academy, XVII, 19–28
Posted 27 September 2018. © British Academy 2018

MARY HESSE
on the occasion of her being awarded an Honorary Doctorate by the University of Cambridge in 2002

Mary Brenda Hesse was born in Reigate, Surrey, on 15 October 1924.[1]

Following a wartime course in electronics and work on the building of transmission receivers, Mary Hesse studied at Imperial College of Science and Technology, London, proceeding to a BSc in special mathematics in 1945, an MSc in 1946 and a PhD in electron microscopy in 1948. She then studied history and philosophy of science under the supervision of Herbert Dingle at University College London, gaining a second MSc in 1950. From 1947 to 1951 Hesse taught mathematics at the women-only Royal Holloway College London, then from 1951 at the University of Leeds. From 1955, following the retirement of Herbert Dingle, she took over the teaching of history and philosophy of science at University College London. In 1960, she was appointed to a university assistant lectureship in philosophy of science at Cambridge, then promoted in 1962 to a lectureship, in 1968 to a readership and in 1975 to a professorship. In 1965, she became one of the founding fellows of the newly formed postgraduate Wolfson College, Cambridge, of which she served as Vice-President from 1976 to 1980. She was elected a Fellow of the British Academy in 1971. In 2002, she was awarded an Honorary Doctorate by the University of Cambridge.

Mary Hesse was active and effective in promoting the recently formed discipline of history and philosophy of science, playing a major role in the 1972 establishment of an independent Cambridge University department in the subject. She served as Vice-President of the British Society for the Philosophy of Science (1970–1), Vice-President of the British Society for the History of Science (1975–7), President of the Philosophy of Science Association (1979–80), on the Council of the British Academy (1979–82), and on the University Grants Committee (1980–5). She edited the *British Journal for the Philosophy of Science* from 1965 to 1969, and from 1975 she was first co-editor then a consulting editor of *Studies in History and Philosophy of Science*. Hesse was widely welcomed as a visiting professor, at the universities of Yale in 1962, Minnesota in 1966, Chicago in 1968 and Notre Dame in 1970. At Cambridge, from 1977 to 1980 she delivered the Stanton Lectures on Philosophy of Religion, and at Edinburgh in 1983, with Michael Arbib, the Gifford Lectures on Natural Theology.

The bulk of Hesse's major contributions to the history and philosophy of science is to be found in her five books (the last of them co-authored).[2] In 1954, aged thirty

---

[1] For further details of her life and career, see M. Hallberg, 'Hesse, Mary Brenda', in S. Brown (ed.), *Dictionary of Twentieth-Century British Philosophers*, vol. 1 (Bristol, 2005), pp. 406–9; M. Hallberg, 'Revolutions and reconstructions in the philosophy of science: Mary Hesse (1924–2016)', *Journal for General Philosophy of Science*, 48 (2017), 161–71.

[2] For a full bibliography, with reviews and secondary literature on Hesse's life and work, see M. Collodel, 'Website in Honour of Mary Hesse', http://www.collodel.org/hesse (accessed 15 March 2018). On her major contributions to the philosophy of science as represented in articles in the *British Journal for the Philosophy of Science*, see S. French, 'Models and meaning change: a brief introduction to the work of Mary Hesse', *British Journal for the Philosophy of Science*, Special Virtual Issue on the Work of Mary Hesse (2017), https://academic.oup.com/bjps/pages/Mary_Hesse (accessed 15 March 2018).

and lecturing on mathematics at Leeds, Hesse produced *Science and the Human Imagination*, a work of extraordinary richness and originality based on her MSc dissertation at University College London and the lectures she had given there. In this book, which contains the germs of much of the later work for which she is renowned, Hesse challenges on historical and philosophical grounds the view of science as an isolated, disinterested activity showing inexorable progress, arguing that 'the sciences, exemplified here by dynamics and astronomy, have always been closely related to their cultural and religious environment'.[3] The first part of the book is largely historical, emphasising practical craftsmanship and medieval Christianity as roots of the Scientific Revolution, and going on to spell out the subsequent ever-sharper divorce of science from religion. The second part of the work provides a detailed critique of positivistic accounts of the status of scientific theories. There follows her own account of their status as analogies, drawn from a wide range of familiar types of experience and reflecting cultural attitudes and preoccupations. In the epilogue, she concludes that 'the practice of scientific research therefore has room for the creative imagination and for recognition of the transcendent, and is not necessarily an arid and impersonal affair, incapacitating the scientist for life in the world of personal encounter'.[4]

In 1961 there appeared the heftiest of Hesse's books, *Forces and Fields*.[5] Meticulously researched, and ranging from the pre-Socratics to quantum field theory, this is widely acknowledged as a major contribution to the history of science. Though the preface acknowledges indebtedness to Karl Popper, the opening chapter on the logical status of theories and the case studies throughout effectively contest hypothetico-deductive accounts of theory justification of the kinds promoted by Popper and Carl Hempel, demonstrating the plurality of criteria for acceptability of scientific hypotheses—empirical adequacy, support by analogical inference, falsifiability, formal simplicity, universality of scope and so forth. The final chapter considers with striking open-mindedness the claims for action at a distance involved in telepathy and clairvoyance.

Models and Analogies in Science came out in 1963.[6] In this, the shortest of her books, Hesse offers a rigorous development of her ideas on the fundamental roles of metaphorical description and analogical inference in the sciences. The first part is cast as a dialogue between Pierre Duhem, dismissive of models as dispensable psychological aids favoured by the broad but shallow English mind, and Norman Robert

---

[3] M. B. Hesse, *Science and the Human Imagination: Aspects of the History and Logic of Physical Science* (London, 1961), pp. 9–10.
[4] Ibid., p. 161.
[5] M. B. Hesse, *Forces and Fields: the Concept of Action at a Distance in the History of Physics* (London, 1961).
[6] M. B. Hesse, *Models and Analogies in Science* (London, 1963).

Campbell, proponent of essential roles for models in the formation and justification of theories. The second part analyses the types of analogy involved in scientific models, showing how they involve both horizontal analogies, that is, similarities of properties, and vertical analogies, that is, correspondences between causal relations. Hesse emphasises the heuristic role of neutral analogies, those whose validity is as yet unknown, as bases for further investigation. In the third part, she provides logical accounts of support of hypotheses through various types of analogical inference. This has proved to date the most influential of Hesse's books, widely cited not only by philosophers but also in cognitive psychology and linguistics, where the past twenty years have seen an escalation in studies of the cognitive roles of analogy and metaphor.[7]

Hesse's most technical production in the philosophy of science, *The Structure of Scientific Inference*, was published in 1974. In this wide-ranging collection, she starts by building on insights of Duhem and Quine on the theory-laden nature of scientific observations and on the semantic status of theoretical terms. According to her network model, the applications of all predicates of the sciences are dependent on their entrenchment in a network of generalisations; and the modes of entrenchment of all predicates, both the relatively observable and the relatively theoretical, are liable to modification. The heart of the work provides rigorous inductive logical explications of confirmation, generalisation and argument by analogy.[8] In the final chapter, Hesse suggests that her network model supports a form of scientific realism, one that can meet the challenges of underdetermination of theory by data and radical discontinuities in the history of scientific theory. This moderate realism can, she claims, explain the instrumental success of the sciences in terms of accumulation of approximate truths.

In 1980 there appeared *Revolutions and Reconstructions in the Philosophy of Science*, a collection of articles from the previous fifteen years. Chief among the revolutions referred to in the title is the then recent move of many philosophers of science, including herself, away from logical-analytic accounts of scientific method and the status of theories to more naturalistic accounts, grounded in the past and present practices of scientists.[9] One major consequence of this shift that she draws

[7] See, for example, J. M. van der Meer (ed.), 'Focus: articles on Mary Hesse and metaphor', *Philosophical Inquiries*, 3 (2015), 41–181, and the references to her work in S. Maasen and P. Weingart (eds.), *Metaphor and the Dynamics of Knowledge* (London, 2000); note that the index of the latter lumps together under 'Hesse, A.' references to Mary Hesse and to the political economist Albert Hesse.

[8] M. B. Hesse, *The Structure of Scientific Inference* (London, 1974), chs. 3–11.

[9] Hesse reflects on judicious appeal by philosophers of science to the history of science in 'The hunt for scientific reason', in P. D. Asquith and R. N. Giere (eds.), *PSA 1980: Proceedings of the 1980 Biennial Meeting of the Philosophy of Science Association*, vol. 2: *Symposia* (East Lansing, MI, 1981), pp. 3–22.

attention to has been recognition of the extent to which theories are underdetermined by data; and Hesse presents the articles in the volume as her attempts, faced with such underdetermination, 'to steer a course between the extremes of metaphysical realism and relativism'.[10] The first part of the book explores the consequences of this revolution for the historiography of the sciences. It focuses on two drastic alternatives to the discredited 'inductive' histories of scientific progress culminating in current orthodoxies: abstinence from all evaluation, in order to understand past sciences in terms of the thought processes of their ages; and the so-called 'Strong Programme', seeking to explain past science in terms of pursuit of social interests. Both are granted limited approval. In the first case, Hesse fully concedes that historians of science should seek to understand past scientific beliefs in the conceptual settings of their periods, while insisting that an element of evaluation from our present standpoint is inevitable if we are to judge what in the past is to count as science.[11] As for the Strong Programme, far from attacking it head on, Hesse welcomes a watered-down version, endorsing the view that 'rational norms and true beliefs in natural science are just as much explananda of the sociology of science as are non-rationality and error', while rejecting 'social determinism' and insisting that the 'cultural norms' appealed to in social explanations of scientific theory should include 'rational rules' adopted in a society.[12] The second part of the book elaborates on the roles of models and analogies in the sciences, and on the complex interactions of theory with observation. Here we see a notable shift away from her previous moderate realism. Where before Hesse had argued that appeal to experientially based models and other conditions for theoretical coherence suffices to defend a moderate realist view of science against the arguments from underdetermination of theory by data, she here concedes that the plurality of experientially based models undermines her earlier position. By way of reconstruction, she moots how the notion of scientific objectivity might be rescued in the context of a pragmatic account of truth as consensus achieved through mutual interpretation and reasonable dialogue. The final chapter touches on truth in theology; and there she declares her commitment to a Christian theology that would 'address the real conditions of our society'.[13]

In *Revolutions and Reconstructions* and her final book (co-authored with Michael Arbib), *The Construction of Reality*, Hesse engages closely with Jürgen Habermas's *Knowledge and Human Interests* and his postscript to that work.[14] This, incidentally, is

---

[10] M. B. Hesse, *Revolutions and Reconstructions in the Philosophy of Science* (Brighton, 1980), p. xiv.
[11] Ibid., ch. 1.
[12] Ibid., pp. 56–7.
[13] Ibid., p. 252.
[14] J. Habermas, *Knowledge and Human Interests* [1968], trans. J. J. Shapiro (London, 1972); J. Habermas, 'A postscript to *Knowledge and Human Interests*', *Philosophy of the Social Sciences*, 3 (1973), 157–89.

what provoked the entry by Hesse's colleague Hugh Mellor in the satirical *Philosophical Lexicon*: 'Hessean, noun. A kind of sackcloth worn at a habermass by those renouncing hemple mindedness.'[15] Hesse was, in fact, no blind devotee. She calls into question several of Habermas's central positions—notably his commitment to a universal conception of rationality and the sharp distinction he draws between natural sciences, grounded in pursuit of prediction and control, and human sciences, grounded in pursuit of mutual understanding. However, there is much that she adopts. In particular, the pragmatic account of truth that she sketches for all fields of inquiry draws on Habermas's account of truth as consensus achieved through free and reasonable dialogue; and she endorses much of his hermeneutic theory, while insisting on its relevance to the natural as well as the human sciences. Hesse also engages with Habermas's views on critique, the quest for liberation from political and ideological domination; and in this connection she considers religion as emancipatory. Where her *Models and Analogies* and *Structure of Scientific Inference* are exemplary in their meticulous arguments from clear premises to clear conclusions, in *Revolutions and Reconstructions* and her contributions to *Construction of Reality* she conducts open-ended explorations and virtual dialogues, in line with her Habermasian vision of conversation as the proper route to consensus. Readers seeking rigorous advancement of specific doctrines may be frustrated; but those who value novel questions and indications of new lines of inquiry will find these works immensely rewarding.

The Construction of Reality appeared in 1986, a year after Hesse took early retirement. This remarkable work attempts nothing less than to 'reconcile an account of the individual's construction of reality … with an account of the social construction of language, science, ideology, and religion'.[16] Such reconciliation is sought through the development of 'schema theory', which sets out to specify the processes through which stable representations are achieved. Hesse's main contributions are in the chapters devoted to social schemas. Her emphasis is on the values in pursuit of which social consensuses are formed. Religion, for example, is presented as the product of social search for a good life, a life freed from evil through communion with God.[17]

On retirement, Hesse launched herself into a new career in landscape history.[18] In 1985 and 1986, she completed Certificates in Landscape History and Archaeology,

[15] D. Dennett and A. Steglich-Petersen, *The Philosophical Lexicon*, 2008 edition, http://www.philosophicallexicon.com/ (accessed 15 March 2018).
[16] M. A. Arbib and M. B. Hesse, *The Construction of Reality* (Cambridge, 1986), second cover.
[17] Ibid., p. 102.
[18] This account of Mary's post-retirement research is based on 'Philosopher to local historian: Mary Hesse', in D. Wilson and F. Midgley (eds.), *Ringing True: Memories of Wolfson College, Cambridge, 1965–2015* (Cambridge, *The Cambridge Review* Committee, 2015), pp. 89–90. My thanks to Fiona Brown for a copy of this anonymous article and to Susan Oosthuizen for sending me her contribution to it,

and in Local History, at Cambridge University's then Board of Extra-Mural Studies (now Institute of Continuing Education). She was active in the Cambridge Antiquarian Society from 1991 to 2001, serving successively as its Honorary Secretary, President and Vice-President; and in the late 1990s she founded the Landscape and Local History Group, a discussion forum for researchers in and outside the University. Her first publications in landscape history dealt with fields, boundaries and land tenure around the Creake villages in north Norfolk.[19] These were followed by studies of Suffolk, including widely cited articles on the arable exploitation and settlement patterns implied by Domesday Book entries.[20] Hesse was a leading member of the South-West Cambridgeshire Project, a community landscape history enterprise run from the University's Institute of Continuing Education from 1997 to 2007. In this capacity, she worked on the reconstruction of medieval field systems in a number of parishes. That material was published both in journals and in informal reports of the project, which she edited with Susan Oosthuizen.[21] At the same time, she researched the identification of Anglo-Saxon and medieval boundaries and hundred meeting places across the Cambridge region.[22] In her final article, of 2007, she reconstructed the development of Cambridge's medieval East Fields.[23]

What holds together this quite extraordinary range of original research and speculation? To borrow three terms from theology, I suggest that Hesse's work can be seen as *eirenic*, *ecumenical* and *syncretic*. It is *eirenic* in its generous and charitable handling of positions at odds with her own. One instance, already noted, is her conciliatory critique in *Revolutions and Reconstructions* of the Strong Programme in the sociology of science; another is one of her last philosophical articles, 'How to be post-modern without being a feminist', published in 1994, in which she approves of feminist contributions to the history and philosophy of science, while distancing herself from the notion of a distinctive feminist epistemology.[24] In line with this charity

together with a listing of Hesse's publications on landscape history, and for checking and improving my account of Hesse's work in this area.

[19] M. B. Hesse, 'Fields, tracks and boundaries in the Creakes, North Norfolk', *Norfolk Archaeology*, 41 (1992), 305–24; M. B. Hesse, 'Field systems and land tenure in South Creake, Norfolk', *Norfolk Archaeology*, 43 (1998), 79–97.

[20] M. B. Hesse, 'Domesday land measures in Suffolk', *Landscape History*, 22 (2000), 21–36; M. B. Hesse, 'Domesday settlement in Suffolk', *Landscape History*, 25 (2003), 45–57.

[21] M. B. Hesse, 'Field systems in Southwest Cambridgeshire: Abington Pigotts, Litlington and the Mile Ditches', *Proceedings of the Cambridge Antiquarian Society*, 89 (2000), 49–58.

[22] M. B. Hesse, 'The Anglo-Saxon bounds of Littlebury', *Proceedings of the Cambridge Antiquarian Society*, 83 (1995), 129–39; M. B. Hesse, 'The field called "Augey" in Ickleton: an Anglo-Saxon enclosure?', *Proceedings of the Cambridge Antiquarian Society*, 85 (1997), 159–60.

[23] M. B. Hesse, 'The East Fields of Cambridge', *Proceedings of the Cambridge Antiquarian Society*, 96 (2007), 143–60.

[24] M. B. Hesse, 'How to be postmodern without being a feminist', *The Monist*, 77 (1994), 445–61.

in disputation is Hesse's constant modesty and generosity in acknowledging sources and precursors of her own innovative views. Notable examples include: the references to works of Alfred North Whitehead, Herbert Dingle and Stephen Toulmin in *Science and the Human Imagination* in connection with her arguments against the privileging of science as the sole form of knowledge and insight; her acknowledgement in *Models and Analogies in Science* of indebtedness to Norman Robert Campbell and Max Black; her citations of the views of Pierre Duhem and Willard van Orman Quine as precursors to her own network theory of meaning; and throughout her later works the credit given to Jürgen Habermas as a source of inspiration.

Hesse's work is *ecumenical* in its perennial concern to overcome false and damaging dichotomies. In her earliest and latest works, she seeks to disarm the conflict between science and religion, reason and faith. As I found in conversations with her, she disapproved of the opposition between Anglo-American analytic philosophy and so-called 'Continental philosophy'. This took courage in Cambridge, where other philosophers did not always react kindly to people such as Gerd Buchdahl, her colleague in the Department of History and Philosophy of Science, and Mary herself when they used such words as 'hermeneutics' and 'phenomenology'. She opposed the isolation of the human sciences from the natural sciences, insisting that the former have to employ some of the techniques of the latter, for instance in 'dating of archaeological findings, and of manuscripts, and reconstruction of historical events from circumstantial evidence'.[25] Indeed, her own work on agricultural history combines 'scientific' environmental history with human history, manifesting an exemplary combination of sensitivity in interpretation of documents and monuments with scientific rigour in matters of authentication, dating, measuring and mapping. As a historian of science, she regretted the isolation of the discipline from mainstream history; and as a philosopher of science, throughout her works Hesse insisted that worthwhile philosophy of science is not an armchair business, but demands scholarly and expert engagement with the contents and practices of the sciences, past and present.

Mary's *syncretism*, her quest for an overarching structure that would bring together the worlds of everyday experience, poetry, the sciences and religion, is most evident in her first and last books. In *Science and the Human Imagination* this unity is glimpsed through values shared in the practices of science and Christianity; and in her final reflections in *The Construction of Reality* schema theory yields intimations of a transcendent reality, a 'God schema' that grounds all human worlds.[26]

---

[25] Hesse, *Revolutions and Reconstructions*, p. 183.
[26] Hesse, *Science and the Human Imagination*, pp. 161–4; Arbib and Hesse, *The Construction of Reality*, pp. 236–43.

Note on the author: Nicholas Jardine is Emeritus Professor of History and Philosophy of Science, University of Cambridge. He was elected a Fellow of the British Academy in 2004.

Annette Dionne Karmiloff-Smith

18 July 1938 – 19 December 2016

elected Fellow of the British Academy 1993

by

JEFFREY ELMAN

LORRAINE K. TYLER
Fellow of the Academy

MARK H. JOHNSON
Fellow of the Academy

Biographical Memoirs of Fellows of the British Academy, XVII, 29–34
Posted 18 June 2018. © British Academy 2018

ANNETTE KARMILOFF-SMITH

What is the origin of the unique and complex behaviours that our species are capable of? How do nature and nurture interact? Is human cognition—our memory, language, numerical abilities—organised into distinct modules? When developmental disorders occur, do they arise from selective damage to domain-specific modules? These age-old questions have puzzled—and often deeply divided—scientists for hundreds of years. Over the course of more than four decades, Annette Karmiloff-Smith's research provided key insights that challenged the traditional answers, leading to new insights into how genetic and environmental factors interact to give rise both to typical as well as atypical outcomes.

Annette Karmiloff-Smith was born in London in 1938. Her parents were Isaac Smith (a tailor and shop owner) and Doris Ellen Smith (née Findlay, an administrator with the Lea Valley Authority). She attended Edmonton County Grammar School from 1949 to 1954, and the Institut Français, London, from 1954 to 1957. Her childhood was somewhat marred by the family's fluctuating fortunes due to her father's successful tailoring business being lost through his heavy gambling during their summer holidays in Nice. Because she loved languages, Karmiloff-Smith decided to work as a simultaneous translator for the United Nations, based in Geneva. However, this proved to be less intellectually stimulating than she had hoped, and she was annoyed by having to translate points of view that she did not agree with. While browsing in a Geneva bookstore, Karmiloff-Smith recognised the famous child psychologist Jean Piaget from book covers and followed him to audit his lectures. She found his research fascinating, and subsequently studied with him, first gaining a bachelor's degree (1970) and then a doctorate (1977) from the University of Geneva.

Piaget was a founding figure of developmental psychology, and his theories dominated the field during most of the twentieth century. While Karmiloff-Smith respected those theories, she was not afraid of challenging them. Nonetheless, throughout her career Karmiloff-Smith's work was deeply influenced by Piagetian thinking in several ways. Like Piaget, she believed that the key to development was to understand the mechanisms that underlie the trajectory of changes across developmental time which give rise to increasing complexity in behaviour, and she modernised this view by helping to establish the 'neuroconstructivist' approach to human development—perhaps most clearly articulated in the co-authored volume *Rethinking Innateness* (1996).[1] Furthermore, like Piaget, Karmiloff-Smith believed that the study of development is inherently an interdisciplinary enterprise requiring input from many disciplines including philosophy, linguistics, genetics and cognitive and developmental neuroscience—all fields to which she contributed. Finally, like Piaget she

[1] J. L. Elman, E. A. Bates, M. H. Johnson, A. Karmiloff-Smith, D. Parisi and K. Plunkett, *Rethinking Innateness* (Cambridge, MA, 1996).

believed in the scientific value of observation (as a key supplement to experiments); one of her early contributions was to pioneer a radically different research strategy for understanding development, the so-called 'microgenetic' approach. This involved observing developmental change at a fine grain of temporal analysis. The method has now become standard in the field.

While Karmiloff-Smith's thinking was deeply embedded in the Piagetian tradition, she saw it as important to question, and even criticise, aspects of the theories articulated in his many books. It is perhaps unfortunate that at the time the Piagetian School was under attack from a number of prominent 'anti-Piagetians', who demonstrated apparently precocious infant and toddler abilities earlier than Piagetian stage theory had predicted. These attacks led to a closing of ranks within Geneva around defending a precise and literal interpretation of Piaget's statements. As Karmiloff-Smith was a student of Piaget who questioned elements of the theory, it was inevitable that conflict with her mentors would arise. Her filed correspondence reveals angry exchanges over a mysteriously withdrawn conference abstract on the limits of the Piagetian account of language acquisition, a position that Karmiloff-Smith later articulated in the 1979 book based on her PhD thesis—*A Functional Approach to Child Language*.[2] The preface to her 1992 book *Beyond Modularity* alludes to these troubled years and her uncomfortable position straddling the divide between the Piagetian School and its critics.[3] Over more recent years, as Piaget's contribution to the field began to be seen more as foundational, rather than literal gospel, Karmiloff-Smith was welcomed back to Geneva and her visits increased. It is unfortunate that her terminal illness prevented her delivering a lecture at a major tribute conference to Piaget in the summer of 2016. Her correspondence and working notes from this period are now lodged with the Piaget Archives in Geneva.

While Karmiloff-Smith's unique position within developmental psychology may have been awkward both with her mentors and 'anti-Piagetians' alike, for those outside the immediate field this was an attraction as it offered a potential synthesis. After brief appointments at the Max-Planck Institute for Psycholinguistics, Nijmegen, and the University of Sussex, in 1982 Karmiloff-Smith found a natural home at the newly created Medical Research Council Cognitive Development Unit in London led by John Morton. Morton encouraged a diversity of views, often debated in heated Tuesday morning theoretical discussions. On the closure of the Unit in 1998, she moved to become Head of her own Neurocognitive Development Unit at the Institute

---

[2] A. Karmiloff-Smith, *A Functional Approach to Child Language: a Study of Determiners and Reference* (Cambridge, 1979).
[3] A. Karmiloff-Smith, *Beyond Modularity: a Developmental Perspective on Cognitive Science* (Cambridge, MA, 1992).

of Child Health in London, during which she strengthened further her interests in neurodevelopmental disorders. After her official retirement from this role in 2003, she enjoyed equally productive years as a Professorial Research Fellow at the Centre for Brain and Cognitive Development, Birkbeck, University of London, a position she held until her death in December 2016.

Karmiloff-Smith's early work was in the area of language development, bringing her into contact with a wide range of scientists—including linguists, cognitive psychologists and developmentalists—who were proposing that the human cognitive system was organised into modules that were potentially innately present from birth. This proposal was hotly debated within all these fields. Karmiloff-Smith entered the debate and proposed what was a fresh and compelling theory that on the one hand did not require innate modules, but which on the other accounted for behaviours that seemed to implicate modular organisation. These ideas were the core of her book *Beyond Modularity*. Karmiloff-Smith explained how modularisation—as a process— might result over the course of development as a result of internal cognitive changes that yielded successively more refined and more modularised knowledge representations (a process she called 'representational redescription'). Thus, modularisation of knowledge need not be innate, but instead is a result of learning and development.

In more recent years, Karmiloff-Smith turned her attention to developmental disorders, both because of their public health importance and as potential windows into the mechanisms that underlie human cognitive development in typical as well as atypical populations. She argued that developmental disorders should not be understood as 'normal minus something broken', but as developmental trajectories that take very different paths from the typical. When one sees what appears to be the same behaviour in both typical and atypical populations, that behaviour may actually be supported by processes that are quite different in the atypically developing individuals than in the typical cases. Karmiloff-Smith's work in this area involved individuals with Down's syndrome, Fragile X syndrome and Williams syndrome, among others. Her findings dramatically reshaped our thinking about both typical and atypical cognitive development.

Most recently, Karmiloff-Smith moved into an area that appears at first to be an odd focus for a developmentalist: Alzheimer's Disease (AD). She pointed to an intriguing phenomenon: by later age most individuals with Down's syndrome have brains that (on autopsy) have the signature characteristics of AD, even though not all of them have the cognitive deficits typical of this form of dementia. The question she asked is what protective factor in these individuals inhibits the behavioural and cognitive effects that are otherwise associated with the brain characteristics of AD. At the time of her death, she was pursuing a fascinating set of hypotheses, working with a broad team of neurologists, geneticists, gerontologists and other specialists to address this question.

Karmiloff-Smith was a generous scientist, mentoring many students and postdocs and inspiring young academics with her zest for science. She was accessible, warm and funny and wonderfully loyal and supportive to her many friends. She was also an excellent communicator, writing books for the general public (*Everything Your Baby Would Ask, If Only He or She Could Talk*),[4] giving talks at festivals (Sunday Times Festival of Education, 2015) and appearing on radio and TV (*The Life Scientific*, BBC Radio 4). She also received many honours over the course of her lifetime. She was elected a Fellow of the British Academy (1993), Academy of Medical Sciences (1999) and Academia Europa (1991). In 2002, she was the first female to receive the European Latsis Prize, in addition to a Doctorat Honoris Causa from the University of Louvain, and in 2004 she was made a CBE in recognition of her contributions to cognitive development.

Annette Karmiloff-Smith served on the British Academy's Communications and Activities Committee and its successor, the Events and Prizes Committee, from 2007 to 2013. In 2001, she gave a lecture at the British Academy to mark the centenary of the British Psychological Society,[5] which turned out to be the first in the highly successful series of annual lectures that have been jointly organised by the two bodies ever since.

Her first marriage to economist Igor Karmiloff, with whom she had two daughters, ended in divorce. She subsequently had a long and happy partnership with Mark Johnson, a professor of cognitive neuroscience, whom she married in 2001. She is survived by her husband, her daughters, Yara and Kyra Karmiloff, brothers Stephen, Peter and Paul, and grandchildren Alexander, Nicholas, Misha, Tatyana, Dylan, Joden and Liliana.

Annette Dionne Karmiloff-Smith, developmental cognitive neuroscientist, born 18 July 1938, died 19 December 2016.

Note on the authors: Jeffrey Elman is Distinguished Professor of Cognitive Science at the University of California, San Diego. Lorraine K. Tyler is Professor of Cognitive Neuroscience at the University of Cambridge; she was elected a Fellow of the British Academy in 1995. Mark H. Johnson is Professor of Experimental Psychology (1931) and Head of the Department of Psychology at the University of Cambridge; he was elected a Fellow of the British Academy in 2011.

[4] K. Karmiloff and A. Karmiloff-Smith, *Everything Your Baby Would Ask, If Only He or She Could Talk* (London, 1998).
[5] A. Karmiloff-Smith, 'Elementary, my dear Watson, the clue is in the genes . . . or is it?', *Proceedings of the British Academy*, 117 (2002), 525–43.

Peter Mathias

10 January 1928 – 1 March 2016

elected Fellow of the British Academy 1977

by

MAXINE BERG
Fellow of the Academy

Biographical Memoirs of Fellows of the British Academy, XVII, 35–50
Posted 27 September 2018. © British Academy 2018

PETER MATHIAS
at All Souls College, Oxford, July 1984

Peter Mathias was Chichele Professor of Economic History at the University of Oxford, the leading chair in economic history, during the period of the subject's greatest strength, from the 1960s to the 1980s. Economic history later went in different directions; some would say that it declined, at least for a time. In his reflections on this later period, Mathias said that economic history lost the battle, but won the war. It had become a central part of mainstream historical research and teaching. If it did so this was in no small part due to Mathias's activities. Richard Smith called him the most internationally oriented of economic historians, and indeed he was fundamental to a major internationalisation of the subject in the third quarter of the twentieth century. His contribution was intellectual, through his focus on comparative industrialisation, and institutional, through his considerable success in building networks that crossed disciplines, countries and barriers of class and culture. He was always open to, and in touch with, new intellectual developments in economic history; he had a quick sense of where the opportunities opened for the field and of the way in which his subject fitted into the wider historical disciplines. He worked on many international and national committees, demonstrating great integrity, courage and fairness.[1]

Peter Mathias was an economic historian of industry, business and technology. His outlook was comparative and he was open to new social science methods. He kept a balance between empirical economic and social history as practised by earlier historians and arising out of the tradition of T. S. Ashton, and the emerging 'new economic history', based in economic theory and quantitative methodologies. He wrote a major monograph, the best-known textbook of his generation, which was translated into many languages. He also published several shorter monographs and many edited collections and articles. More than any other economic historian of his generation, Mathias contributed to making his field both internationalist and broadly conceived, of appeal to both social scientists and historians.

---

[1] See the obituary notice by J. A. Davis, 'Peter Mathias CBE 1928–2016', *Journal of European Economic History*, 45 (2016), 1–8. Also see extended interviews with Peter Mathias: Negley Harte and Peter Mathias, Economic History Society Interviews with Economic Historians, 2002, LSE Archives, https://archives.lse.ac.uk/Record.aspx?src=CalmView.Catalog&id=ECONOMIC+HISTORY+SOCIETY%2f2010%2f12%2f25&pos=26 (accessed 14 March 2018); A. McFarlane, 'An interview on the life and work of the historian, Peter Mathias, Part 2, 23 September 2009' http://www.repository.cam.ac.uk/handle/1810/226099 (accessed 14 March 2018); and A. McFarlane 'Interview with Peter Mathias', 18 December 2008 www.alanmacfarlane.com/ancestors/mathias.htm (accessed 26 March 2018). See, in addition, P. O'Brien, 'Preface', in K. Bruland and P. O'Brien (eds.), *From Family Firms to Corporate Capitalism: Essays in Business and Industrial History in Honour of Peter Mathias* (Oxford, 1998), pp. vii–xi, and 'Peter Mathias: a bibliography', in Bruland and O'Brien, *From Family Firms*, pp. xix–xxv. I have also benefited from unpublished obituary notices by Richard Smith and Negley Harte.

I

Peter Mathias came from a modest rural background. His family originated in Dublin, but his father, Jack Mathias, was born in Plymouth; he went on to work in the Plymouth dockyards and later joined the Royal Navy. His mother Marion Love was from the small Wiltshire village of Wingfield. Many in her family had been in service to the Caillard family at Wingfield House. Peter was born on 10 January 1928 in Freshford, Somerset, and spent his early years there with his mother and grandparents, while his father was on active duty abroad in the Navy. In 1932, the family moved to Bristol where his father became a clerk in the local building society. A teacher spotted Peter as a bright boy, and in 1938 entered him for a scholarship to Colston's Hospital, a local charity and direct grant grammar school. He boarded there and, while his father was again away on naval duties during the Second World War, his mother returned to her parents in Wingfield. Peter spent his school holidays with them and did much of his growing up in these rural surroundings. He remembered a community of tiny workers' cottages with large families and great poverty. Extra tutoring and some classes at Bristol Grammar School helped him in his application to Cambridge where he gained an exhibition at Jesus College in 1946. He did his National Service before taking up his college place in 1948.

Mathias's tutor at Jesus was the medievalist, Vivian Fisher, while Charles Wilson taught him English Economic History. Wilson was an enormous influence, joined, in his later undergraduate years, by Sir Michael (Munia) Postan. A further mentor in his early undergraduate years was Edward Welbourne, Senior Tutor and later Master of Emmanuel College. Peter graduated with a distinction in the History Tripos in 1951.

Mathias went to Welbourne's lectures during all three years, and later described him as an eccentric, very traditional about Cambridge, but radical in economic history. Welbourne had written on the Durham miners, but was an embattled figure of the right in Fabian-dominated economic history at the time. It was Wilson, a leading figure on English mercantilism and Anglo-Dutch relations, who sent Mathias to Welbourne's lectures. Wilson too took a strongly empirical and anti-Fabian or anti-socialist approach to his subject, and he also fostered an academic approach to business history. Wilson, at the time Mathias came to Cambridge, was already involved in the great history of Unilever, along with a team of research assistants in London and Rotterdam. This was to be the first major academic business history in Britain, resulting in a two-volume work in 1954.

Wilson became Mathias's research supervisor and arranged for him to go to Harvard for some economics training but he forgot to send in the application, so this was delayed for a year. Mathias spent 1952–3 at Harvard, taking part there in the early years of the Research Center for Entrepreneurial History (1948–58). Wilson had good international connections that he would pass on to Mathias, first within the USA, then later with the European University Institute and the Datini Institute. Even more significant for Mathias's trajectory was his early encounter with Postan. Wilson did not like Postan; therefore, as an undergraduate, Mathias had to sneak off to Postan's lectures. Confronted once by Wilson, he defended himself: 'you must admit that he knows a lot about the manor'. Wilson retorted, 'He should do; he grew up on one.'

Postan had come from the London School of Economics (LSE) to the Cambridge Chair of Economic History in 1938 and was the major force in economic history in Cambridge throughout Mathias's time as a student and then lecturer in the Cambridge History Faculty. Postan introduced Mathias to a wider world beyond Cambridge— that of Weber, Dopsch, Durkheim, Marx and wide reading in European economic history. Postan was as much at home among the economists as the historians. He lectured on Marxism for the Economics Faculty and on both medieval and twentieth-century economic history for the History Faculty. A medievalist with an up-to-date knowledge of writing on the twentieth-century economy was a rare thing, but Postan had gained his expertise during his wartime service in the Department of Economic Warfare. He ran the famous Economic History seminar in Cambridge which Mathias attended from 1951. Both Wilson and Postan pushed him, as a young researcher, to take up opportunities and to respond to influences outside Cambridge. Mathias went to W. W. Rostow's lectures in 1951–2 on 'The British Economy of the Nineteenth Century' and, some years later, he attended his series on 'The Stages of Economic Growth'. Apart from his year at Harvard, Mathias was also a regular participant in T. S. Ashton's seminars at the LSE. He remembered Donald Coleman cooking meals after the seminars in his flat in Charing Cross Road where debate continued; many brought wine but this never quite measured up to Coleman's expectations. Mathias later made brief excursions to the École des Hautes Études in Paris. He was centrally involved as a student and young researcher then lecturer in the making of new directions in the field at Cambridge, London and Europe, learning much from European traditions as well as from the USA.

II

Mathias was a research student for only a year before being elected a Research Fellow of Jesus College, Cambridge, whereupon his supervisor, Charles Wilson, told him he no longer had any need for a PhD degree. He was researching the brewing industry, a subject that combined his interests in industrial and business history. This was an important topic, perhaps less appreciated at the time than it is now when historians have recently turned to research the histories of food and drink. Mathias made pioneering use of business records held by many surviving local brewery firms throughout the country, although access to them was seldom easy. The brewers were very conservative and reticent, but Mathias had cousins in the industry, and Welbourne too had links to Dale's Brewery in Cambridge. Mathias researched with great success using records that had suffered the 'hazards of efficient chief clerks, the salvage drives of two world wars, and the Luftwaffe (all destroying records in their different ways and effective in that order)'.[2]

The book Mathias wrote, *The Brewing Industry in England 1700–1830* (1959), set high standards for future industrial and business histories, and remains unsurpassed as a study of brewing.[3] The volume also conveyed the significance of the industry to historians of the factory system. It demonstrated that the transition to porter brewing had been part of an early and very rapid shift from small-scale and household processing to very large-scale and mass production; this was thus a frequently neglected case of industrialisation. The book led naturally to further interests connected with those of Charles Wilson, arising out of Wilson's Unilever study. Mathias went on to write *The Retailing Revolution: a History of Multiple Retailing in the Food Trades Based upon the Allied Suppliers Group of Companies* (1967).[4] This new book formed an important contribution to academic business history.

Fast on the heels of this book came another that more truly conveyed Mathias's central interest in the history of industrialisation in Britain: his textbook, *The First Industrial Nation: an Economic History of Britain 1700–1914* (1969),[5] which went through many imprints, with the last in 2001. It remains in print. This work was based on the lectures Mathias gave in Cambridge from 1955 onwards when he took over the first-year economic history course from Charles Wilson. The book marked the end of his long teaching on this course. In 1968, aged forty-one, he left Cambridge to take up the Chichele Chair in Economic History at All Souls College, Oxford.

The First Industrial Nation became the main text alongside Hobsbawm's *Industry and Empire* during the years when economic history grew to its height in the historical disciplines.[6] It was read in sixth forms and universities across Britain and the English-speaking world, then translated into many languages; it made Mathias one of the best-known economic historians. Its lucid coverage of most of the key topics, combining

---

[2] N. Harte, 'Peter Mathias', Obituary Notice.
[3] P. Mathias, *The Brewing Industry in England 1700–1830* (Cambridge, 1959).
[4] P. Mathias, *The Retailing Revolution: a History of Multiple Retailing in the Food Trades Based upon the Allied Suppliers Group of Companies* (London, 1967).
[5] P. Mathias, *The First Industrial Nation: an Economic History of Britain 1700–1914* (London, 1969).
[6] E. J. Hobsbawm, *Industry and Empire: the Pelican Economic History of Britain*, vol. 3: *From 1750 to the Present Day* (Harmondsworth, 1969).

accounts of individuals with basic economic concepts and a good set of statistical tables and graphs, complemented another key text of the period, Phyllis Deane's *The First Industrial Revolution* (1965), which was designed around economic concepts and data.[7] Mathias made use of basic concepts of economic theory and time-series data in the way of Clapham and Ashton; indeed, the book was dedicated to 'T.S.A.'. The chapters were short and lively, and it became *the* book for students of the subject in its heyday, with economic growth as its theme and sound historical explanations of how modern economic growth had begun in Britain.

III

Mathias's inaugural lecture of 1970 in the Chichele chair (which he had taken over from H. J. Habakkuk) was entitled 'Living with our Neighbours'. It responded to the impacts of the new sociology on social history and of economic theory on economic history.[8] Mathias pointed to the impetus provided to social and economic history by the new study of social relations as a dynamic in the process of economic growth, especially through demography, law, education and cultural structures of the market. He warned economic historians against retreating into examining exclusively economic variables, and of relying on data and measurement outside of context. In Mathias's view, economic history would 'remain a largely synthesizing discipline' drawing new ideas from the social sciences but 'the limitations of data would condemn it to remain an inexact science': 'The methodological claims of economic history to be a social science are modest, but it can challenge over-ambitious methodologies among economists and sociologists.'[9]

His caveats were to some degree acknowledged by the first wave of new economic historians emerging from the USA from the end of the 1960s, though largely forgotten by the next. And Mathias, in his Chair in Oxford, was instrumental in bringing the new economic history to Britain through visiting fellowships that brought Americans, Europeans and other international scholars to All Souls. He had arrived there at an opportune moment. Not long before, the college was criticised in the Franks Commission of 1964 for 'its infirmity of purpose'. Habakkuk had suggested a visiting fellows programme as a means of avoiding it becoming a college for graduate students, rather than a fellows-only institution. Since the late 1960s, over six hundred visiting

[7] P. Deane, *The First Industrial Revolution* (Cambridge, 1965).
[8] P. Mathias, *Living with our Neighbours: the Role of Economic History: An Inaugural Lecture delivered before the University in the Hall of All Souls College, Oxford, on the 24th of November, 1970* (Oxford, 1971).
[9] Also see the version published as P. Mathias, 'Living with our neighbours', in N. B. Harte (ed.), *The Study of Economic History: Collected Inaugural Lectures, 1893–1970* (London, 1971), pp. 1–20.

fellows have participated, and each year between twelve and twenty are elected. Mathias had the contacts and the energy to implement the scheme and he quickly found a number of key players to fill the slots.

The visiting fellowships gave Mathias the opportunity to bring in established and ascendant figures in the field in Europe, North America and elsewhere: Herman Van der Wee, François Crouzet, David Landes and Rondo Cameron, Jacob Price, Richard Goldthwaite and Louis Cullen, and the new wave of quantitative economic historians, Robert Fogel, Albert Fishlow, Robert Galman, Peter Temin and Paul David were all visiting fellows over the next twenty years. Nuffield College, with Max Hartwell as Reader in Economic History there, brought in other visiting economic historians, including Nathan Rosenberg and Stanley Engerman, and had studentships and post-doctoral fellowships to offer, which brought those working in economic history and economics alongside the new social history. Max Hartwell assiduously built up the economic history holdings at the Nuffield Library, quickly acquiring the newest journals, working papers and books, especially those from the USA. At All Souls, Mathias ran the celebrated Friday evening seminar in economic history, crowded out with those from several faculties.

While not becoming himself a 'new economic historian', Mathias brought his knowledge of industry and technology in the eighteenth century to bear on debates during the 1970s on the role of technology and innovation in economic growth among the new economic historians, and also took part in debates on the role of science and technology arising out of the new social history of science. He brought several histor-ians of science to All Souls just at the key juncture of these debates and himself published several important articles on these subjects, some of which were gathered into his *Science and Society 1600–1800* (1972) and later into *The Transformation of England* (1979).[10]

During these Oxford years, Mathias was also the editor of the distinguished multi-volume *Cambridge Economic History of Europe*, its first volume edited by J. H. Clapham and Eileen Power, and several subsequent volumes by Postan. Mathias took on the final volumes and brought the enterprise to a successful completion.[11] He also started and was series editor of the very influential Methuen series, 'Debates in

[10] P. Mathias (ed.), *Science and Society 1600–1900* (Cambridge, 1972); P. Mathias (ed.), *The Transformation of England* (London, 1979).
[11] P. Mathias and M. M. Postan (eds.), *The Cambridge Economic History of Europe*, vol. 7, parts 1 and 2: *The Industrial Economies: Capital, Labour and Enterprise* (Cambridge, 1978), and P. Mathias and S. Pollard (eds.), *The Cambridge Economic History of Europe*, vol. 8: *The Industrial Economies: the Development of Economic and Social Policies* (Cambridge, 1988).

Economic History', one that did much to shape the teaching of the subject over more than two decades.[12]

IV

Equally important as his writing and teaching, and his position in the Chichele chair, were Mathias's roles in the Economic History Society and, connected with this, in the International Economic History Association (IEHA). As a final year undergraduate, he was taken from Cambridge to his first meeting of the Economic History Society by E. E. Rich. This was the first residential conference of the Society, at Worcester College, Oxford, in 1951. It was so cold that R. H. Tawney came down to breakfast wrapped in a blanket. He lit his pipe during the first session, and proceeded to set this blanket on fire, creating quite a scene. Tawney's great commitment to keeping subscriptions to the Society and the *Economic History Review* low was later continued by Mathias, together with Kenneth Berrill, both of whom transformed the Society's finances when they became successive Treasurers, making shrewd investments, increasing membership and initiating larger print runs of the journal to bring costs down.

The Oxford conference was the beginning of the residential conferences of the Society, and Mathias quickly became a driving force in the Society, working along-side Habakkuk, Postan and Berrill. Berrill, the bursar of St Catharine's and later King's College, Cambridge, had worked in the City, and invested the Society's funds in Japan and in a Dutch investment company. He once carried gold in a belt to Japan to invest for the Society. Mathias continued the tradition, deftly investing so that he was able to raise the Society from a small organisation during the 1930s relying on 'whip rounds' in Eileen Power's kitchen to the best-funded subject group organisation in the humanities and social sciences by the 1980s.

Mathias's work for the Society, as Reviews Editor, then as Assistant Editor of the *Review* from 1955, Treasurer of the Society (1968–88) and President (1989–92) made it a formative institution for his own and other research students. Experiences for these students were rich and varied. There was a great frisson, with debates of the moment between economic and social historians, catching of late-night buses to sleep on the floors of various friends near to conferences held across the North of England, and going off on conference excursions where Mathias and his friends and colleagues on the Council led the way to mills, sites of former open field agriculture, ports and customs houses.

---

[12] A full listing of his publications is 'Peter Mathias: a bibliography', pp. xix–xxv

Mathias, both within Oxford and within the Economic History Society, fostered a broad and connective economic history. While bringing so much of the new economic history to Britain and Europe, he viewed the subject as one also connected with social and cultural history. And indeed, the Economic History Society grew as a society of economic and social history. The period when Mathias presided over economic history was also the time of the great rise of social history with History Workshop and the new Social History Society, along with the great period of labour history. Mathias's 'broad church' kept the fields together, whereas in the USA they divided.

V

The Economic History Society and Mathias's early close link to this through Postan also brought him into the origins of the IEHA. Both Berrill and Mathias assisted Postan in the early meetings, most held in Paris with Fernand Braudel. The IEHA grew out of panels and meetings at the CISH (Comité International des Sciences Historiques) Congress in Stockholm in 1960. Meetings were held beforehand in Paris, and in Stockholm with Ernst Söderlund, then Professor of Economic History at the University of Stockholm.

Mathias went to many of these, though not funded for his travel, and reported 'turning out his pockets' to find the cash for his expenses. It was not easy dealing with Braudel, but Braudel had access to the funding, some of this provided through the Rockefeller and Ford Foundations as well as the French government. Braudel also had many close connections with young East European scholars through the visiting fellowships he could offer them, and the raison d'être of the IEHA was to provide an opening and access to Western scholarship for historians, especially of the smaller East European countries. A small group of mostly likeminded Europeans formed the Executive Committee, and Mathias was on this from the start. The Committee was as much an arm of international diplomacy as it was a group for organising international conferences. It brought together West and East Europeans, the Americans and the Russians in this effort in its early days. Mathias called the IEHA a 'Cold War project', and he was pivotal in negotiating the complexities of various stages of détente while on the Committee, especially in his periods as Secretary (1959–62) and when he was President between 1974 and 1978.[13]

Mathias played a particularly important role over Czechoslovakia after it was invaded by the Warsaw Pact in 1968. The Czech economic historian, Arnošt Klima,

[13] See M. Berg, 'East–West dialogues: economic history, the Cold War and détente', *Journal of Modern History*, 87 (2015), 36–71.

was on the Executive Committee of the IEHA and was well known in the West for his publications on the transition from feudalism to capitalism and later on proto-industrialisation in *Past & Present* and the *Economic History Review*. His student, Jan Palach, set himself on fire in Wenceslas Square at a political protest on 16 January 1969. Klima was summoned to the Czech Academy of Sciences and asked whether he approved of the Russian intervention. His negative answer sealed his fate, and he was no longer allowed to attend international meetings outside the Iron Curtain. Mathias stood up to the pressure from the Soviets and others in the Eastern bloc to have Klima removed from the Executive Committee. At the Edinburgh Congress in 1978, when Mathias was President of the IEHA, the Soviet delegate, Vinogradov, made a further effort to have Klima's now conspicuously vacant seat filled by another Czech representative. Mathias had the decision deferred, and phoned Klima that evening; miraculously his call went through, and he learned that membership of this prestigious international committee was the only thing keeping Klima out of jail.

Mathias concluded the business: 'I reported this to the Committee in Edinburgh. Vinogradov remained silent, his gun having been spiked, and Klima was duly re-elected.' Mathias followed up with a sharp letter to the Minister of Education in Prague, protesting the denial of a passport for Klima, which had caused 'widespread dismay and anger amongst economic historians in many countries'. Klima remained on the Committee, his chair vacant until he resigned in 1982. He was replaced by the Hungarian Iván Berend, though only after an objection by the Russians that another Czech should be elected was overruled.

Klima had been in the USA in 1968 and had participated in the International Congress at Bloomington along with his compatriots, Alice Teichová and Mikuláš Teich, who were also on American visiting fellowships at the time and were invited speakers. Together with Klima they were the only Czechs at the Congress, though it was also attended by a few Poles and some Hungarians including György Ránki and Iván Berend. A Soviet presence was a high priority for the central committee of the Communist Party of the Soviet Union, and it sent a strong delegation under the leadership of Vinogradov. The subject of the invasion was a great event at the Congress. Teichová spoke from the floor at a crowded session on industrial structure in the twentieth century; she was applauded and supported by other East Europeans.[14] The Teichs did not return to Czechoslovakia after their fellowship year; Klima did return because his family was still there. Mathias found college fellowships in Cambridge for the Teichs, and supported Teichová's appointment to a lectureship and later chair in economic history at the University of East Anglia.[15]

---

[14] See discussion of these events in Berg, 'East–West dialogues', 62–5.
[15] Interview with Alice Teichová and Mikulus Teich, 23 March 2010.

Mathias remained active in the IEHA as an Honorary President, attending the Congresses right up to 2009 at Utrecht. He was also a member from the start in 1967 of the Datini Institute in Prato initiated by Fernand Braudel and Federigo Melis. Braudel accepted Mathias, while he excluded Postan. Mathias was on the scientific executive committee until he was seventy, and took part in the meetings, often three times a year, to organise an annual week-long conference around different themes of medieval and early modern economic history. Several on the Prato committees were also on the IEHA, and the annual Settimane brought many British economic historians into contact with other Western and Eastern European as well as American and Canadian scholars.

Mathias also had connections with Italy. He was one of the founders of the *Journal of European Economic History* (*JEEH*), working closely with Luigi de Rosa of the Maritime University in Naples. The *JEEH* was funded by the Banco di Roma, and Mathias was active on the international board from its founding in 1972 to 2016. The Bank of Rome wanted a journal focused on Europe in its broadest geographical sense, but also addressed to an international audience; hence, its language of publication was English.[16] Mathias continued to take a close interest in the *JEEH* long after de Rosa's death, and checked and proof-read each issue before it went to press right up until his own death.

Through de Rosa, Mathias also became involved in the enterprises of the Avvocato Gherardo Morrato at the Istituto Italiano per gli Studi Filosofici in Naples. The Istituto launched a series of English-language summer seminars in economic history for Italian students—the first was in Oxford in 1984, led by Mathias. Four further seminars were held at the Centre for Social History at Warwick University, co-directed with Professor John Davis. The papers were published in five volumes edited by Mathias and Davis.[17] Mathias gave annual lectures at the Istituto Filosofici, later published in *Cinque Lezioni di storia e teoria dello sviluppo economico* (2003). His last lecture in Naples was 'L'idea di Europa: Mutamenti di concetti e realtà attraverso i secoli', written in the aftermath of the 2008 financial crisis.[18] In this he looked at changing perceptions of the idea of Europe through to the development of the post-Second World War European project.[19]

---

[16] See P. Mathias, '*The Journal of European Economic History*: a review of its existence over forty years since its foundation in 1972', *Journal of European Economic History*, 40 (2011), 11–27.

[17] See P. Mathias and J. A. Davis (eds.), *The First Industrial Revolutions* (London, 1989); P. Mathias and J. A. Davis (eds.), *Innovation and Technology in Europe* (London, 1991); P. Mathias and J. A. Davis (eds.), *Enterprise and Labour* (London, 1996); P. Mathias and J. A. Davis (eds.), *Agriculture and Industrialization* (London, 1996); P. Mathias and J. A. Davis (eds.), *International Trade and British Economic Growth* (London, 1996).

[18] Published by the Istituto Italiano per gli Studi Filosofici, Naples: La Scuola Pitagora Editrice, 2009.

[19] See Davis, 'Peter Mathias CBE 1928–2016'.

Mathias was also head of the Advisory Board of the Central European University Press, and through this he became acquainted with George Soros, since the Press was part of the Central European University, the institution that Soros had so generously endowed. He had, in addition, close links with the Wissenschaft Collegium in Berlin, with the United Nations Educational, Scientific and Cultural Organization (UNESCO), and held fellowships with Danish and Belgian academies. Mathias's other major international connection was with Japan, where he was an advisor to the Sasakawa Foundation. In 1983, the Japanese Crown Prince, Naruhito, came to study for two years in Oxford and was a member of Merton College with Mathias as his supervisor. Mathias not only supervised but found research assistance for Naruhito's project, and helped with its publication. He also provided great pastoral care, taking Naruhito on his first visit to the archives in the Oxford County Record Office, organising frequent social gatherings with other students and taking him to Ironbridge. He became close to the Japanese royal family and saw Naruhito on frequent visits to Japan over the next two decades. He also built on a link earlier initiated between Downing College, Cambridge, and Keio University to exchange visiting fellows when he went there as Master of Downing in 1987. He added a large annual summer school programme for junior members of Keio at Downing. Mathias became one of two international advisors to Keio, was awarded an honorary degree there and received the award of the Order of the Rising Sun with Gold Rays at the Japanese Embassy in London in 2003.[20]

VI

Naruhito was only the most illustrious of Mathias's many research students. Mathias supervised most of his PhD students in Oxford, but Brian Harrison came to him earlier, while Mathias was still in Cambridge. Harrison, an Oxford student, sought him out because he wanted to write on the temperance movement, and a man who knew about brewing would be able to help. As Oxford Professor of Economic History, Mathias also had many international students. Others from Japan included Heita Kawakatsu, now Governor of Shizuoka Prefecture, and there were also many from various parts of Europe, Australia, Canada and the USA, as well as British students. The subject was at its height, and at some points Mathias had over fifteen PhD students. These he supervised with individual care, quickly identifying those who needed

[20] I am grateful to Professors Toshio Kusamitsu and Takao Matsumara for some of these details of Peter Mathias's connections with Japan. Also see Prince Naruhito, *The Thames and I: a Memoir of Two Years at Oxford* (Folkestone, 2006: English Translation of *Thames no tomo ni*, 1993).

more expert supervision and arranging this, making connections with publishers, and writing the carefully crafted references that saw many of his students into academic jobs or helped them in the pursuit of other career paths. His students also took part in the Economic History Seminar, held in the All Souls Old Library, and other faculty history, politics and society seminars, and for some the economic and social history meetings run by Max Hartwell in Nuffield. It was an exciting time to be a research student in Oxford, with easy access also to all that was going on in London, as it had been for Mathias when he was a student in Cambridge, and as it continued there in different areas of the field. Mathias took a genuine personal interest in his students' lives and entertained them at home and in college. Though not an easy person for younger people to talk to, he had a loyal following, his former research students staying in close touch long after their PhDs. In one case, he attended the ordination and later installation of his former student Canon Dr Edmund Newell.

Significantly, these atmospheres, intellectual conjunctures and social networks have to be made; and those who reside mainly in their studies are not the ones who make the subject. Mathias lectured regularly and was a major figure on many university committees, chairing the large Oxford History Faculty for a period. All of Mathias's research students gained from the contributions he made to his subject, where, in Oxford, it lived with its neighbours in social and radical history, economics and sociology. In the very slowly changing course structures of Oxford in the 1970s and 1980s, Mathias, along with Patrick O'Brien, introduced a new cross-faculty degree in History and Economics. Moreover, Mathias never saw himself as leading a 'school' of history. He was open to the subjects brought in, and developed by, his students; indeed, he expected his students to find their own topics, unlike a number of his predecessors and even colleagues.

Towards the end of his years in Oxford, many of the separate economic and social history departments across the country were struggling to survive, and nearly all eventually merged with wider history departments, or some of their members joined economics faculties. Mathias accepted this; he had never been keen on artificial boundaries and thought horizons broadened when these departments eventually merged with history. In his view, this was the way the 'war' of the subject areas was won. Economic history was now fully a part of wider historical studies.

VII

Mathias left Oxford in 1987 when he was elected Master of Downing College, Cambridge, a post he held until 1995. He had no previous connection with the college, and succeeded the charismatic John Butterfield, long a college member before his election as Master. Mathias came in as an outsider, despite his early formation in Cambridge. The transition was not easy; he lost the opportunity to supervise research students too soon in his career, as he had no Faculty appointment. This was certainly a loss, but he continued to participate when he could in the Economic History Seminar in Cambridge, then led by the Chair of Economic History, Barry Supple. He was soon to be brought into the centre of university administration as the Vice-Chancellor's deputy on many syndicates and committees but was not a part of the History Faculty life of Cambridge. He also came from a rich college without students to a modestly endowed institution with a large undergraduate community and a strong sporting ethos. The college chaplain, Bruce Kinsey, described the difference: 'All Souls was "civilized"; the view from the window was of someone raking the gravel [...] this contrasted hugely with the hurly-burly of Downing often more hearty than arty [...] and a boat club, night climbers and rugby club that was often noisy, heavy-drinking and reckless. Mathias was magnificent on the history of brewing, but less on the experience and consumption of beer.'[21] Mathias tackled the college finances as he had done those of the Economic History Society. He had inherited the first stages of a project for the new Howard Building, and with the support of Alan Howard saw it through to completion, as well as that of Howard Court. He raised further funding to build the Maitland Robinson Library. Together with his wife, Ann, he made the lodge at Downing a centre of entertainment.

Mathias was elected to a Fellowship of the British Academy in 1977, serving as its Treasurer from 1980 to 1989 during a period of great expansion in the Academy's funding due to an increase in the government grant-in-aid, and in the market value of investments, as well as the new devolution of the management of Postgraduate Funding in the Humanities to the Academy. His period as Treasurer was also during the planning of, and move to, new premises, from smaller shared spaces in Burlington House to the much larger building on Cornwall Terrace. He chaired the Academy's Records of Social and Economic History Committee (1990–6); and he took a special interest in the overseas British schools and institutes, an important part of the Academy. He was also an 'independent member' of the Advisory Board for the Research Councils, and thus contributed on a broad basis to research funding policy. He was awarded a CBE in 1984. He made up for his lack of a PhD by being made DLitt by Oxford in 1985 and by Cambridge in 1987, and became an Hon. DLitt at least six other universities, including the University of Warwick. Among his roles on many advisory boards he was President of the Business Archives Council, and was particularly supportive of the Modern Records Centre at the University of Warwick, which collected business and labour records. He was instrumental in the choice of the

---

[21] Rev. Bruce Kinsey, 'Oration at the Funeral of Peter Mathias'.

Modern Records Centre for the records of BP, and the addition to the library building that came with this.

Mathias married Ann Blackmore in Bath Abbey in 1958. They made family homes in Cambridge and Oxford, and lived after retirement first in Bassingbourn Mill, Cambridgeshire, then in Chesterton. They were both collectors, Mathias of a remarkable set of eighteenth-century trade tokens, and both of Newhall and other porcelain.[22] They had three children, Sam, Sophie and Henry, and often took the family to sites of agricultural and industrial history as well as to their holiday house in Norfolk, where they enjoyed the local wildlife. Mathias remained close to his roots, with frequent family visits to Wingfield, the small village in Wiltshire where he had been a boy. In Mathias's later years he continued, together with his wife Ann, his travels to Japan and also to Italy, until his increasing physical immobility and then her death confined him closer to home. He kept up with his subject and received many visitors from around the world in his field and outside it. He died on 1 March 2016 and his ashes are buried with those of his parents in Wingfield Parish Church in Wiltshire.

Acknowledgements

I worked with Peter Mathias, first as a research student between 1972 and 1976, then as a research fellow until 1978, and afterwards in the Economic History Society, the IEHA and the Datini Institute. I have gathered information for this memoir from my own interviews and conversations with him in the early 2000s and in the years before his death, as well as from the obituary notices and interviews listed in note 1. I have also spoken and corresponded with several of his colleagues and associates and former students, including, in particular, Kristine Bruland, John Davis, Marguerite Dupree, Anne Hardy, Brian Harrison, Negley Harte, Edgar Jones, Bruce Kinsey, Toshio Kusamitsu and Takao Matsumara, Edmund Newell, Patrick O'Brien, Barry Supple, Rick Trainor, Herman van der Wee and John Wood. I am grateful to several of these and to Sophie and Henry Mathias, Pat Hudson and Peter Brown for reading an earlier draft.

Note on the author: Maxine Berg is Professor of History at the University of Warwick. She was elected a Fellow of the British Academy in 2004.

[22] P. Mathias, *English Trade Tokens: the Industrial Revolution Illustrated* (London, 1962).

Brian Benjamin Shefton

11 August 1919 – 25 January 2012

elected Fellow of the British Academy 1985

by

JOHN BOARDMAN

Fellow of the Academy

ANDREW PARKIN

Biographical Memoirs of Fellows of the British Academy, XVII, 51–61
Posted 18 June 2018. © British Academy 2018

BRIAN SHEFTON

Brian Shefton was a highly regarded classical archaeologist whose wide-ranging contributions to his discipline were internationally recognised. He spent most of his academic career at Newcastle University in the north-east of England, but was a truly international scholar as well as an inveterate traveller, devoting a great deal of his time to research trips and attendance at academic conferences. He retained a remarkable commitment to his academic work throughout his life, continuing to be an active and engaged scholar long after his retirement in 1984.

Brian Benjamin Shefton was born Bruno Benjamin Scheftelowitz on 11 August 1919 in Cologne, Germany. He was the younger son of Isidor Isaac Scheftelowitz, Professor of Indo-Iranian Philology at the University of Cologne and a rabbi, and his wife, Friedericke (Frieda), née Kohn. His education began at the Apostelgymnasium, a school linked to the Roman Catholic Apostelkirche in Cologne. However, in the summer of 1933 the Scheftelowitz family were compelled to move to Britain to escape Nazi oppression. The family first settled in Ramsgate where Isidor Scheftelowitz secured a teaching position at Montefiore College, while Bruno became a pupil at St Lawrence College. In the summer of 1934, they moved to Oxford where Isidor had been invited to lecture on Zoroastrianism to the Faculty Board of Oriental Languages and Literature. Much of that period involved lengthy commuting between Ramsgate and Oxford, which precipitated Isidor Scheftelowitz's kidney condition. After Isidor's death from kidney failure in December 1934 the family remained in Oxford and Bruno, who had been enrolled at Magdalen College School, went on to read classics at Oriel College in 1938.

During his time at Oxford, Bruno became drawn towards the study of Greek archaeology and attended classes given by Paul Jacobsthal and (Sir) John Beazley, both of whom were to become major influences on his subsequent academic work. He recalled that he was inspired to study Greek archaeology by an encounter in the British Museum with a red-figure volute krater by the Altamura painter that depicted a Gigantomachy.[1] This led him to seek out Beazley and ask to be admitted to his special course on Greek archaeology, despite missing the official date for enrolment. Beazley was persuaded to admit him, obtaining permission for Bruno to take his special subject, and at the same time took him on as a personal pupil, initiating a relationship that was to continue until Beazley's death.

Bruno's studies were interrupted by the outbreak of the Second World War. In 1939, he helped to pack up the Greek vases in the Ashmolean Museum, and in the summer of 1940 he was interned, alongside many other refugees from Europe, on the Isle of Man.[2] His internment, a period about which he never talked with his immediate

---

[1] Vicky Donaldson interview with Brian Shefton for her PhD research.
[2] There has been some doubt about Shefton's internment, but his daughter Penny has drawn attention to a number of family papers that refer to this.

family, was relatively short-lived and after his release he enlisted in the British army in October 1940. He carried out his training at a Pioneer Corps centre in Ilfracombe, Devon, alongside other German and Austrian nationals. After training, he initially served in 249 (Alien) Company Pioneer Corps and was involved in camp construction in Scotland. In November 1944, he transferred to the Royal Army Education Corps. At some point in his military career, he decided to anglicise his name to Brian Benjamin Shefton. Once the war was over, he resumed his studies at Oxford and graduated with a First Class degree in 1947.

After Oxford, Shefton was elected to a School Studentship by the British School at Athens in 1947. Subsequently, he held the Derby Scholarship from Oxford, and was awarded a Bishop Fraser Scholarship from Oriel College. These awards allowed him to spend three years in Greece and Turkey. This extended period immersed in the archaeology of the region led to a number of publications and had a deep impact on much of his ensuing work. He became involved in various different projects, including the British School's excavations at Old Smyrna where he assisted Richard V. Nicholls (a future Keeper of the Fitzwilliam Museum), in particular with understanding the site's fortifications. He also became involved in the excavations of the American School of Classical Studies at Athens in the Athenian Agora, working on some of the pottery finds. He helped identify, for example, some Clazomenian pottery and added some fragments to a krater attributed to the Kleophrades painter. He also produced a detailed study of a louterion from the Agora later published in *Hesperia*.[3] A further area of interest was the pottery from Perachora, excavated by Humfry Payne for the British School at Athens before the war. He wrote a chapter on the imported pottery found at Perachora in one of the publications produced for the site.[4] In addition, he carried out research on the dedication of Kallimachos from the Epigraphic Museum in Athens, which led to an article and short postscript in the *Annual of the British School at Athens*.[5]

During his time at the British School, Shefton travelled extensively in Greece despite the dangers of making journeys in a country that was in the midst of a civil war. He was allegedly the first student after the Second World War to walk from Olympia to the temple of Apollo at Bassai. He remembered frequently hearing the

---

[3] B. B. Shefton, 'Herakles and Theseus on a red-figured louterion', *Hesperia*, 31 (1962), 330–68.
[4] See B. B. Shefton, 'Other non-Corinthian vases', in A. A. A. Blakeway, T. J. Dunbabin and H. G. G. Payne (eds.), *Perachora: the Sanctuaries of Hera Akraia and Limenia: Excavations of the British School of Archaeology at Athens, 1930–1933. 2: Pottery, Ivories, Scarabs and other Objects from the Votive Deposit of Hera Limenia* (Oxford, 1962), pp. 368–88.
[5] B. B. Shefton, 'The dedication of Callimachus (IG 12 609)', *Annual of the British School at Athens*, 45 (1950), 140–64; B. B. Shefton, 'The dedication of Callimachus. A postscript', *Annual of the British School at Athens*, 47 (1952), 278.

sounds of gunfire in the mountains during this expedition, as well as being fearful of the fierce wild dogs that inhabited this part of Greece, although neither was enough to put him off his goal of visiting the temple. On another occasion, he managed to persuade the Greek navy to give him a lift so that he could visit the monasteries on Mount Athos. This enthusiasm for travel, as well as a determination to overcome any difficulties to get to where he wanted to go, was to remain a prominent feature of his life right up until his final years.

On his return to Britain in 1950, Shefton was appointed as a lecturer in classics at the University College of the South West of England in Exeter. Here he worked on Greek material in the Royal Albert Memorial Museum in Exeter and on other material in the University's collection. He also published an important attribution study of three Laconian pot-painters.[6]

In 1955, Shefton moved to Newcastle upon Tyne to take up a post as a lecturer in ancient history at King's College, which was to become the University of Newcastle upon Tyne in 1963. Here he was encouraged by the vice-chancellor, Charles Bosanquet, to create a small collection of classical antiquities to support the teaching of Greek archaeology.[7] Bosanquet, son of the British archaeologist Robert Carr Bosanquet and born in Athens during his father's directorship of the British School, wanted to build up Greek archaeology in Newcastle; Roman archaeology was already well established. Starting from a small initial grant of £25, Shefton developed the collection and, through a combination of support from Newcastle University, the National Art Collections Fund and bequests and gifts from external benefactors, on his retirement in 1984 it consisted of nearly one thousand objects. He was tireless in his work to set up and ensure a secure future for the collection and rightly considered it one of his finest achievements. Indeed, he often stressed the fact that the collection was as much a research resource as one for teaching, and he was frequently drawn to objects that he thought had research potential. He became a familiar figure in the salerooms of London as he sought out pieces, operating across the worlds of academia and the art market in ways that subsequent university archaeologists would not be able to do. Auction houses, such as Sotheby's, as well as private collectors, including George Ortiz, came to rely on his opinion about many objects that came onto the market. At times his enthusiasm got the better of him, such as the occasion, at a Sotheby's auction, when he had to be told that he was bidding against himself. It was not only through the salerooms that he managed to add to the collection. He also relied on the support

---

[6] B. B. Shefton, 'Three Laconian vase-painters', *Annual of the British School at Athens*, 49 (1954), 299–310.

[7] J. Boardman, A. Parkin and S. Waite (eds.), *On the Fascination of Objects: Greek and Etruscan Art in the Shefton Collection* (Oxford and Philadelphia, PA, 2016) provides an overview of the collection, as well as some more detailed studies of specific objects.

of a number of benefactors who gave or loaned material. The most significant of these was Lionel Jacobson, a prominent Newcastle businessman, who was the chairman of the clothing firm Burton. He also received backing from many other individuals and funding bodies such as the London Hellenic Society and Tyne Tees Television. Perhaps the most remarkable acquisition came from the collection of Dr Leo Mildenberg in Switzerland.[8] Shefton had noticed the similarity between half a terracotta lion's head in Newcastle and a similar half from the Mildenberg collection. They were both halves of a terracotta water spout from the site of San Biagio close to the Greek colony of Metapontum in southern Italy. The fact that these were two halves of the same object was confirmed in 1981 when a cast of the break in the Mildenberg fragment was taken and matched with the Newcastle half. Mildenberg generously bequeathed his half of the lion's head to Newcastle and it arrived, amidst much publicity, in 2004.

Over the years, the collection moved around the university campus. An initial display of Greek material was put on in the University's Hatton Gallery, where Shefton, in association with Ralph Holland, an art history lecturer at Newcastle and collector in his own right, organised an exhibition in 1956. By the 1960s, the collection had relocated to a more permanent home, alongside the Classics Department, in the Percy Building, where it was called the Greek Museum. In the 1990s, when Classics moved into the Armstrong Building, the collection travelled with it. A museum space was created in an area that had formerly been chemistry laboratories, and in 1994 this was named the Shefton Museum of Greek Art and Archaeology to honour Shefton's outstanding contribution to classical archaeology at Newcastle University. More recently, the collection was transferred to the Great North Museum: Hancock as part of a major development which saw Newcastle University concentrate its archaeology collections into a refurbished Hancock Museum. The Hancock was traditionally a natural history museum, but a £26 million redevelopment saw natural science collections displayed alongside archaeology and ethnography. In 2010, to mark Shefton's ninetieth year, the Greek Gallery in the Great North Museum: Hancock was renamed the Shefton Gallery of Greek and Etruscan Archaeology.

Alongside the collection of artefacts, Shefton was keen to develop Newcastle University's library into a first-class research centre for Greek and Etruscan archaeology. He applied himself to acquiring books for the University library with the same energy and enthusiasm that he devoted to the acquisition of archaeological material, often bringing volumes back from his travels. The University library was persuaded to support his book buying and eventually a distinct Shefton section was created to

---

[8] B. B. Shefton, 'A Greek lionhead in Newcastle and Zurich', *Antiquity*, 59 (1985), 42–4.

house the results. He conceived of the library as a vital adjunct to the Greek and Etruscan archaeology collection, enhancing its usefulness as a research resource.

During his time at Newcastle University, Shefton's energies were mainly taken up with developing, consolidating and promoting the archaeology collection, as well as with teaching. He surveyed a selection of what in his view were some of the most interesting pieces from the collection in *Archaeological Reports*, a survey that provides an insight into the thinking behind certain acquisitions.[9] In the 1970s, he also set up a team of job creation placements and volunteers to assist with the conservation, display and documentation of the collection. A product of this period was the publication in 1978 of a booklet entitled *Greek Arms and Armour*, which highlighted one particular strength of the collection.[10] Nevertheless, despite all this collection-based activity, Shefton still managed to be actively engaged with research.

Over the course of his time in Newcastle, the University recognised Shefton's important contribution to his discipline, promoting him to Senior Lecturer in 1960, Reader in 1974 and ultimately to a personal Chair in Greek Archaeology in 1979. After a long and distinguished career at Newcastle, Shefton's retirement from his teaching post in 1984 opened up a whole new chapter in his life. He saw retirement as an opportunity to turn his attention wholeheartedly towards research and would publish a number of scholarly articles, many of which highlighted his interest in the distribution of Greek and Etruscan material culture in Iron Age Europe.[11] These articles made use of his remarkable knowledge of museum collections throughout the world, bringing together diverse strands of evidence to inform his arguments. At the same time, he maintained a strong commitment to attending conferences and delivering lectures in the UK and abroad. He continued to be an effective lecturer into his nineties. Tony Spawforth recalled a lecture delivered at the British Museum when Shefton was ninety, where he stood throughout, despite his hip problems, and spoke without notes.[12] His depth of knowledge and interest for his subject were never more apparent than when he spoke in front of an audience.

Shefton's academic career was distinguished by his almost limitless curiosity and broad range of interests. One of his major undertakings was the translation and, to a large extent, rewriting of Arias and Hirmer's *A History of Greek Vase Painting*.[13] Another notable achievement was the publication of a monograph on Rhodian

---

[9] B. B. Shefton, 'The Greek Museum, University of Newcastle upon Tyne', *Archaeological Reports*, 16 (1969–1970), 52–62.

[10] P. Foster, *Greek Arms and Armour* (Newcastle upon Tyne, 1978).

[11] A select bibliography of his publications up to 2004 can be found in K. Lomas (ed.), *Greek Identity in the Western Mediterranean: Papers in Honour of Brian Shefton* (Leiden, 2004), pp. xx–xxii.

[12] A. J. S. Spawforth, 'Introduction', in Boardman, Parkin and Waite, *On the Fascination of Objects*, p. 3.

[13] P. E. Arias, M. Hirmer and B. B. Shefton, *A History of Greek Vase Painting* (London, 1962).

Bronze oinochoai.[14] These two publications highlight a shift that took place in his research focus as he moved from mainly producing studies of painted pottery towards an interest in the metalwork of the Greeks and Etruscans. His early career was largely devoted to Greek vase studies, where his approach was often informed by his studies with Beazley, but he soon moved beyond this, in particular with research on bronze vessels and the archaeology and art of the western Mediterranean. Here his debt to Paul Jacobsthal is more apparent, as they shared an interest in Greek and Etruscan imports into Iron Age Europe.

His vase studies, exemplified by the English edition of *A History of Greek Vase Painting*, covered the traditional fields of iconography but were by no means confined to the usual Attic repertory. He worked on Laconian vase-painters, on shapes, and characteristically on vases which, from their form and decoration, held information about the non-Greek too. This concern for the interaction of the Greek and the non-Greek encompassed the influence of Persian metal shapes, as well as the range of influence of Greek shapes and patterns (the 'Castulo cups', Etruscans and 'eye-cups') and techniques. Sculpture was by no means ignored, with his studies both technical and iconographic.

Evidence for the traffic in pottery led Shefton into studies of trade in general over the whole Mediterranean, especially to the west and in Europe, to the north of the Alps, but also to the east in the Achaemenid Persian period, notably on homeland Phoenician sites. Increasingly, he became as much at home with the non-Greek as the Greek. His study of bronze vessels took him into research on Central European products and the Celts. All this represented a wider range of interests than was common in a classical archaeologist of the post-war years, and much remains relevant to modern studies by scholars who share his diverse interests but, possibly, not his depth of understanding of a wide variety of subjects. His publications are notable for their documentation, footnotes often outweighing text.

Shefton's academic achievements were marked by a number of honours and important fellowships. He was elected Fellow of the Society of Antiquaries in 1980 and Fellow of the British Academy in 1985. The British Academy awarded him their Kenyon Medal in 1999, in a ceremony at Newcastle University that took place during a conference in his honour.[15] He held various significant fellowships, including Leverhulme, Getty at Malibu, Balsdon at the British School at Rome and British Academy with the Israel Academy, while the early 1980s saw him spend a fruitful

---

[14] B. B. Shefton, *Die 'Rhodischen' Bronzekannen*. Marburger Studien zur Vor-under Frühgeschichte, vol. 2 (Mainz, 1979).

[15] Lomas, *Greek Identity in the Western Mediterranean*.

period in Vienna. He was especially gratified by the honorary doctorate awarded by his father's former university, Cologne, in 1989.

Despite his continual travelling from 1955 onwards, Shefton was firmly based in Newcastle. In 1960 he married Jutta Ebel, a Swedish national, and they set up home in the Jesmond area of the city, where they brought up their daughter Penny, who was born in 1963. Jutta Shefton, a translator, played an important part in Shefton's career, supporting his academic work, particularly through her knowledge of several European languages and her skill as a typist. She was instrumental, for example, in the publication of *A History of Greek Vase Painting*, typing up the final manuscript on the kitchen table of the family home. She also provided Shefton with a stable home environment to which he could return after his numerous trips abroad. Shefton was a well-known figure in Jesmond and a keen member of the residents' association. At the same time, he took an active part in the life of Newcastle's small Jewish community.

Shefton's was a larger-than-life personality and most people who met him would not forget the encounter. His curiosity was insatiable. In a library, he would discover what each student was working on, discuss it with them, giving and receiving in equal measure. His camera was a major instrument of recording and research, used freely in libraries, museums, irrespective of any regulations, and even during lectures, where he could often be seen photographing the screen to capture the speaker's slides. The end result was a massive archive of negatives, which sat alongside his equally impressive personal library and paper archive. In fact, his home became so full of books and papers that he had to build an extension to accommodate the ever-increasing mass of material.

Brian Shefton was a truly international scholar, building up wide-ranging networks of contacts throughout Europe, North America and beyond. These networks informed a great deal of his research, which relied on an intimate knowledge of archaeology collections and their curators in many countries. His ability as a linguist meant he could work effectively in several European languages and this was another means by which he created fruitful associations with international colleagues. He also made frequent use of the telephone to keep in touch with his numerous contacts throughout the world and was an enthusiastic convert to email and other digital technologies when they came along.[16] He could also be very persuasive and often managed to get people to go out of their way to help him. For instance, in the last years of his life

---

[16] Shefton's daughter Penny recalled numerous late-night phone conversations with various luminaries of classical archaeology, including Elke Böhr, Konrad Schauenberg, Jean-Jacques Maffre and Dietrich von Bothmer. In fact, she felt he spent most of his research life living abroad whether he was physically in the UK or not.

when he found walking difficult, he convinced an Athenian bus driver to divert significantly from his bus route to drop him at the main entrance of his hotel.

He was one of the first Western scholars to visit Albania, during the dictatorship of the early 1970s, and had numerous other adventures, including a brush with the authorities in Marshal Tito's Yugoslavia where he spent some time locked in the back of a police van. This commitment to overseas travel is illustrated by his attendance at a conference in Basel on Greek pottery north of Etruria in October 2011, just three months before he died.[17] By this time, his mobility was severely impaired and he was reliant on two walking sticks to get about. Nevertheless, he travelled on his own from Newcastle to Edinburgh, where he caught a flight to Cologne. After spending some time in Cologne he then took a train to Basel to attend the conference. This was typical of Shefton, who refused to let any physical problems derail his plans and who would resolutely refuse any help when it was offered. He frequently pointed out that if he started to accept help he would become reliant on it, and this would curtail his ability to travel and participate in the academic world that meant so much to him.

Sir John Burn, Professor of Clinical Genetics at Newcastle University, provided one of the best estimations of Shefton's career when investing him with an honorary fellowship in 2005: 'When it comes to the stuff of which a university is made, there's nothing like a steady, predictable member of staff and Brian Shefton was and is nothing like a steady, predictable member of staff. Rather he is the stuff of what great academic institutions are built: imaginative, bold and irrepressible.'[18]

Acknowledgements
This memoir of Brian Shefton has greatly benefited from several accounts of his life produced by other scholars, not least David Gill's entry 'Shefton, Brian Benjamin (1919–2012)', *Oxford Dictionary of National Biography* (https://doi.org/10.1093/ref:odnb/104851, accessed 14 March 2018) and his chapter 'Brian Shefton: classical archaeologist' in S. Crawford, K. Ulmshneider and J. Elsner (eds.), *Ark of Civilization: Refugee Scholars and Oxford University 1930–1945* (Oxford, 2017), pp. 151–60. The introduction to K. Lomas (ed.), *Greek Identity in the Western Mediterranean: Papers in Honour of Brian Shefton* (Leiden, 2004) has provided further useful content. The memoir has also drawn on personal reminiscences of Brian Shefton from colleagues and friends, particularly Tony Spawforth and Sally Waite, as well as his daughter Penny.

[17] S. Bonomi and M. A. Guggisberg (eds.), *Griechische Keramik nördlich von Etrurien: Mediterrane Importe und archäologischer Kontext* (Wiesbaden, 2015). This volume of conference proceedings was dedicated to Shefton in recognition of his valuable contribution to the study of Greek and Etruscan imports into northern Europe.

[18] Sir John Burn, quoted in *The Times*, 1 March 2012, 47.

Note on the authors: Sir John Boardman is Lincoln Professor Emeritus of Classical Archaeology and Art, University of Oxford; he was elected a Fellow of the British Academy in 1969. Andrew Parkin is Keeper of Archaeology at the Great North Museum, Newcastle upon Tyne.

Charles Peter Brand

7 February 1923 – 4 November 2016

elected Fellow of the British Academy 1990

by

MARTIN MCLAUGHLIN

Biographical Memoirs of Fellows of the British Academy, XVII, 63–70
Posted 27 September 2018. © British Academy 2018

PETER BRAND

Peter Brand, as he preferred to be known, was born in Cambridge on 7 February 1923. His father was a printer and his mother a secretary who also typed up theses for students in the town, so he was not born into privilege. However, after attending the Cambridgeshire High School for Boys, his precocious intelligence won him an Open Major Scholarship to Trinity Hall, Cambridge, where he began a degree in Modern Languages in 1941. Like many of his generation, his university course was interrupted in 1943 by the intensifying conflict of the Second World War, but his language skills were useful to the army and he spent the end of the war going round Sicily and Southern Italy on a motorbike investigating Fascists and Fascist sympathisers. In peacetime he returned to Cambridge and completed his BA in 1948, obtaining a first-class degree in Italian, Spanish and Portuguese. It was around this time that he met and married his Swedish wife, Gunvor.

After his BA, Brand went on to do his PhD under the then Professor of Italian, E. R. Vincent, on 'The Italianate fashion in early nineteenth-century England', receiving his doctorate in 1952. His teaching career began with a short spell at the Perse School in Cambridge and he also taught briefly as an assistant in the Italian department at Edinburgh University: although he was captivated by the Scottish capital, he did not get on well with the head of department, the eccentric Mario Manlio Rossi, and soon returned to a post as University Lecturer in Cambridge and Fellow in his old college, Trinity Hall. However, when in 1966 the Edinburgh Chair became vacant (Rossi retired and returned to Italy), Brand successfully applied for the job (perhaps Edinburgh was ready for a change from the unconventional Italian whose main research interest was in esoteric philosophy). Brand went on to hold the Edinburgh chair with great distinction until retirement in 1988, presiding over the expansion of his department as it became one of the foremost units of Italian studies in the UK. The success of this department was almost entirely due to Brand himself: he was a highly talented administrator and, given his skills in this area, it was no surprise that he went on to become Dean of the Faculty of Arts, which allowed him to recruit replacement staff for Italian. His administrative career culminated in him becoming Vice Principal of Edinburgh University in 1984–8. It was a triumphal and well-deserved end to his academic life.

However, devoted as he was to his beloved Edinburgh and Scotland, Brand did not confine himself to promoting Italian studies north of the border. He was very much a 'good citizen' at national level. He was the Honorary Treasurer of the main subject association, the Society for Italian Studies (SIS), from 1957 to 1962. Here too he was a force for change: one of the reforms he instigated was to persuade the SIS to publish, in addition to its academic journal *Italian Studies*, an annual newsletter, which later became *The Bulletin of the Society for Italian Studies*, with an emphasis on practical developments in teaching. He then went on to serve on the editorial board of the main journal of the SIS, *Italian Studies* (volumes 31–6, for the years 1976–81). However, his reforms were not always welcomed: he had a difficult relationship with J. H. Whitfield, who became Chair of the SIS in 1962, in succession to Brand's former supervisor, E. R. Vincent. The initiative to launch the *Bulletin*, for instance, did not have the Chair's approval. Relations remained frosty and Whitfield published quite a harsh review (in *Italian Studies*, 21 (1966), 122–4) of Brand's 1965 monograph on Torquato Tasso, though this was countered by more generous reviews elsewhere. Brand continued to maintain a national profile, serving as General Editor of the *Modern Language Review* for volumes 66–73 (1971–8), as well as acting as subject Editor for Italian and Spanish in the journal. His national contribution did not go unnoticed: he was elected a Fellow of the British Academy in 1990, and in 1995 became President of the Modern Humanities Research Association. He was recognised beyond the UK as well: in 1970 he was invited as Visiting Professor to Cornell University in the USA, and he also held visiting professorships in Italy, a country that made him a Cavaliere dell'Ordine al Merito della Repubblica Italiana in 1975, and then in 1988 a Commendatore.

Brand's skills as teacher and head of department were of the same high order as his more widely known achievements as a scholar. He was an inspirational teacher who lectured with great clarity and charisma, and his practical, 'can-do' attitude genuinely helped students who were sometimes daunted by taking up a new language and literature: he would regularly explain to undergraduates how they too could work a forty-hour week in an effective and feasible way. His administrative brilliance was evident first and foremost in his own department, where all colleagues were treated with exemplary fairness. His willingness to take on major administrative roles also benefited his own department. When in the late 1970s he agreed to become Dean of the Faculty of Arts, he did so on condition that the Faculty grant him a replacement for a departing colleague and an extra lectureship to help cope with the rising numbers of undergraduates in Italian: indeed, in that year the only two appointments in the whole of humanities at Edinburgh were in Italian, surely a unique occurrence in any modern-day university. Brand also had a justified reputation for making deals with colleagues in his own and other faculties, and significantly many of these deals were struck during or after a tennis or squash match, or over lunch in the university's staff club. In return for all this, the Italian departmental team had to do more than perform well in teaching and research: they also had to become involved in the social activities that helped make it a cohesive unit. Colleagues were encouraged to join in the annual reading party at a country house in Angus owned by the Scottish universities, an event to which students and staff from other Scottish Italian departments also willingly came. With one of the earliest language assistants at Edinburgh, Laura Caretti, a dynamic personality, Brand launched the idea of an annual Italian play, to be put on in Edinburgh by both undergraduates and staff every spring. The plays performed ranged from Machiavelli's *La mandragola* to works by De Filippo and Dario Fo, and they attracted not just a university audience (which included coachloads of students from Glasgow and Aberdeen) but also large numbers from the Italian community in Edinburgh – all this long before Outreach and Impact were coined in their current academic meanings.

Brand himself recalled that when he took over the Edinburgh chair in 1966 there were only two lecturers in the department, and dangerously low levels of undergraduates; but by the end of the 1970s there were four full-time lecturers, and student enrolments were so high that the Faculty had to put a cap on the numbers taking Italian. This was, he said, 'one of the most exciting and cheering periods of my life'. Part of the secret of this success was his campaign to have the beginners' course in Italian accepted as a graduating course that would count towards the degree. There was some hostility to this idea in the faculty, but he succeeded in convincing one of the main opponents, the professor of French, after a drink with him at the university Staff Club following a game of tennis. This was typical of Brand: he took great pleasure in sporting and social activities for themselves but he always had an eye to securing practical advantages. He was a networker before his time, and it is no surprise that some of his colleagues in other faculties nicknamed him 'Brando' (after the lead actor in the two films of *The Godfather*, which were very popular in the early 1970s). But he never rested on his laurels or neglected his own department. Thanks to conversations with Gunvor, he developed a new small-group teaching structure for the crucial beginners' course, which was based on the Swedish car manufacturer SAAB's idea that small units were much better than lengthy conveyor belts.

In terms of scholarship, Peter Brand will be remembered for three major monographs as well as his generous editorial work for *Modern Language Review*. His first book, based on his Cambridge thesis, was *Italy and the English Romantics: the Italianate Fashion in Early Nineteenth-Century England* (Cambridge, 1957). This was a substantial, fifteen-chapter, interdisciplinary study of what he terms the 'Italomania' that seized England in the first forty years of the nineteenth century. It was not just a literary study. The four main divisions of the book give a good idea of its scope: travel and language; the influence of Italian literature in England; the arts and landscape; history, politics and religion. To round it off there is a substantial conclusion, eight illustrations, and eight appendices outlining the huge range of sources he drew on for the volume. Given this breadth it is no wonder that Brand's first monograph is still a point of reference for scholars today, sixty years after its first publication.

Brand's second book saw him move from the nineteenth century to the Renaissance period, but there was still an English dimension. In *Torquato Tasso: a Study of the Poet and of his Contribution to English Literature* (Cambridge, 1965) he devoted seven of the ten chapters to Tasso's life and works, and three to the poet's reception in England. This was a milestone volume in that it was the first complete study in English of Tasso, and it also provided a comprehensive survey of the poet's influence and reputation in England up to the end of the nineteenth century. One particularly strik-ing aspect of the book is that in the part devoted to Tasso's life and works there was no bibliography in English for Brand to draw on: it was genuinely a pioneering vol-ume. The book contains some sensitive literary analyses, particularly of the epic *Gerusalemme liberata* and the pastoral drama *Aminta*. In the substantial forty-page chapter on the *Liberata*, the line of enquiry is largely thematic, but Brand does reserve a quarter of the chapter for an assessment of Tasso's language and style in the poem, with plenty of quotations to illustrate his points. His down-to-earth approach, unburdened by the Crocean jargon that had dominated Italian studies of Tasso, or by modern theory, made the book indispensable for students studying this difficult author.

His final monograph was a book devoted solely to the other great writer of Renaissance chivalric epic, Ariosto, and this time there was to be little attention to the poet's critical fortune in English. *Ariosto: a Preface to the 'Orlando furioso'* (Edinburgh, 1974) was also pioneering in that it was the first of the twelve volumes that would be published by Edinburgh University Press in a series devised by Brand himself, 'Writers of Italy'. The aim of each volume in this series, as he states in the preface, is 'to explain and interpret the work of a major Italian writer'; the books are 'designed also to be intelligible and of interest to those whose knowledge of Italian is slight' (p. vii). That is why each quotation is accompanied by an English translation, unlike in the book on Tasso, and with a view to this broader readership there are no footnotes, though there is a good, up-to-date bibliography at the end. Of the ten chapters, there are two on Ariosto's life and 'minor works', then seven devoted to the thematics and style of the epic, and one on the poem's general reception, with just a few pages on the poem's critical fortune in the UK. As in the volume on Tasso, there are helpful thematic accounts (in the three chapters on love, arms and politics) and sensitive poetic and narratological analyses (in the chapters on narrative, poetry and language). Some of the conclusions hint at why Brand found Ariosto so congenial: towards the end of the chapter on politics, he writes: 'Out of the loves and encounters of his knights and damsels come not so much answers for the ills of his society, but meditation and ten-tative counsel on the business of living with our fellows, friends, masters, enemies, daughters, wives' (p. 125). The author of the Ariosto book always found a way of living humanely with those around him, be they colleagues, friends or family.

This talent for bringing the values he found in Renaissance works into his life was particularly evident in the case of another favourite text, Castiglione's *Book of the Courtier*. Not surprisingly, this work appeared in the Writers of Italy series, when Brand commissioned John Woodhouse to write what became one of the most successful of its volumes: *Baldesar Castiglione: a Reassessment of The Courtier* (Edinburgh, 1978). Castiglione's most famous concept was that of *sprezzatura*, a neologism coined by the author to mean a kind of outward nonchalance which somehow accompanies brilliant achievements. Brand loved teaching *The Courtier* to undergraduates and explaining its relevance, and there was something of that seemingly effortless excellence in many areas of his own life.

Among his other publications were two coedited volumes, which bookend his career looking both backwards and forwards: one was a Festschrift for his supervisor, coedited with Kenelm Foster and Uberto Limentani, *Italian Studies presented to E. R. Vincent on his Retirement from the Chair of Italian at Cambridge* (Cambridge, 1962); the other was the substantial and very successful history of Italian literature coedited with Lino Pertile, Brand's successor at Edinburgh: *The Cambridge History of Italian Literature* (Cambridge, 1996; revised paperback edition, 1999). There was one final book of translations, cowritten with Richard Andrews and Corinna Salvadori: *Overture to the Opera: Italian Pastoral Drama in the Renaissance: Poliziano's* Orfeo *and Tasso's* Aminta *with Facing English Verse Translations* (Dublin, 2013). For this work, Brand translated Tasso's *Aminta*, a play that he had read and loved when he first studied it on an Italian course with Army Intelligence in the 1940s and which he finally translated for the last volume he contributed to.

Peter Brand died on 4 November 2016, and with his passing British academia lost the most successful promoter of Italian studies in Scotland. In these days of small nationalisms, it is worth noting that the person who did most to promote Italian studies in Scotland in recent times was an Englishman who had fallen in love with the country (he was also a fanatical supporter of the Scottish rugby team). In broad academic terms, his legacy will be his three major monographs, as well as the twelve volumes of the Writers of Italy series and the monumental *Cambridge History of Italian Literature*.

Outside academia, Brand loved sport and the outdoors, and would often escape with family, friends or colleagues to his caravan, first in the borders, then at Blair Atholl. His wife of over sixty years, Gunvor, died suddenly in 2010, but Brand coped with his grief with typical fortitude. A number of colleagues from different parts of the UK and Ireland had the pleasure of attending his ninetieth birthday party at a restaurant in Edinburgh in February 2013. The next day he generously invited all of them to a lunch at his house, but only after he had taken his daily morning walk up Corstorphine Hill. His infectious love of life stayed with him to the end. Perhaps that quality also came from one of his favourite authors. In his Ariosto book, while speaking of the 'message' of the *Furioso*, he notes: 'Such philosophy as emerges, is a personal one, a human, humane and common sense acceptance of love as a power for good or evil in men's lives' (p. 59). The human, humane and common sense are all qualities that were epitomised by Peter Brand himself. Very much a family man, he is survived by three daughters Jane, Anne and Catharine, and a son, Simon.

Acknowledgements

I am indebted to a number of people for some of the points in the above memoir: Phil Cooke, Peter France, David Robey and Peter Brand's children. A useful summary of all of Brand's published works up to 2000 can be found in Martin McLaughlin (ed.), *Britain and Italy from Romanticism to Modernism: a Festschrift for Peter Brand* (Oxford, 2000), pp. 178–82.

Note on the author: Martin L. McLaughlin was Professor of Italian and Agnelli-Serena Professor of Italian Studies at the University of Oxford; he won the British Academy's Serena Medal in 2017.

Malcolm Beckwith Parkes

26 June 1930 – 10 May 2013

elected Fellow of the British Academy 1993

by

VINCENT GILLESPIE
Fellow of the Academy

Biographical Memoirs of Fellows of the British Academy, XVII, 71–87
Posted 27 September 2018. © British Academy 2018

MALCOLM PARKES

Malcolm Parkes, DLitt, FBA, died on 10 May 2013 at the age of eighty-two, after a long illness.

It is a cold, foggy day in December 1971. I am sitting in Stephen Wall's room in Keble College, Oxford, immediately after lunch. The gas fire is hissing, Stephen is making small talk and we are waiting for the arrival of the other admissions interviewer. All of a sudden the door flies open and in sweeps a middle-aged man of medium height and portly build, surrounded by a miasma of damp tweed and pipe smoke. 'Sorry', he says to Stephen, 'telephone'—as if that explained everything. (Later in life I would come to realise it did, given the usual duration of his phone calls; the phrases 'make it quick' or 'to cut a long story short' were an infallible sign that you were in for the long haul, requiring cancellation of the newspapers and sending out for emergency rations.) Positioning himself in front of the gas fire, and fixing me with an intense look he recites:

> Puffs, powders, patches, bibles, billet-doux.

'What effect does the word "bibles" have in that line?' he asks. By sheer luck (and good teaching at secondary school) I recognise it as being from Pope's *The Rape of the Lock*, and burble on about p sounds and b sounds, and about incongruity. But he wants more, keeps pushing me to say something about the punctuation, about the progression and juxtaposition of the ideas, about the metre of the line. With great kindness but dogged persistence, he refuses to accept the first answer: 'Go on', he says after each comment, encouraging me to challenge and develop the initial formulaic response. After what feels like a lifetime of flailing, we move on to other things, and my first intellectual encounter with Malcolm Parkes (hereafter MBP) is over. But I have often thought of it since as indicative of all that was great about him both as teacher and scholar—the intense interest in the sounds the words make; in the way they are deployed on the page; in their punctuation and their meaning in the wider context of the text itself. All these were part of what made MBP not just a great and innovative palaeographer but also a superb cultural historian: he always wanted to look beyond *how* the scribe was copying the text to ask *why* he was doing so. He read and thought about the copied texts as well as assessing and classifying the hands used to copy them. This skill as a highly attentive close reader was a formidable reinforcement of MBP's visual acuity, the meticulous and remarkably retentive memory for hands that allowed him to engage with the palaeography and codicology of manuscript books written from the seventh through to the seventeenth centuries. And it made him an inspiring undergraduate tutor, a challenging but supportive graduate supervisor and a hugely entertaining, bracing and occasionally exasperating colleague.

MBP used often to refer with approval to G. W. Prothero's *A Memoir of Henry Bradshaw* (London, 1888) as an affectionate account of one of the greatest English palaeographers, codicologists and book historians of the nineteenth century.[1] What Prothero says of Bradshaw in his preface could equally well be applied to Parkes himself, and I have often wondered if it served as some sort of inspiration for his own work and teaching, or if he recognised something of himself in the portrait:

> Over those with whom he came in much contact ... his intense individuality, with its strange blending of strength and tenderness, of frankness and sensibility, of human affection and scientific enthusiasm, exercised an irresistible fascination. The width and exactitude of his knowledge, the thoroughness of his research, his elevation of science above all thought of self, his respect for genuine study in all branches however remote from his own, gave to many students a new ideal and a stimulus all the more potent because it was suggested rather than enforced.

What is certain is that one can apply Prothero's praise of Bradshaw's influence on and guidance of colleagues and students precisely to MBP's impact on his field:

> The help which he so ungrudgingly gave is acknowledged in many grateful prefaces and recorded in many learned notes, but such indications attract little attention and the original worker is easily forgotten or ignored.

That should not be the fate of MBP's indirect contributions to scholarship, as his many friends and admirers continue to value his wisdom, cite his datings and rely on his opinions. Meanwhile, his direct contributions to scholarship through publication are embedded in the core bibliography of the many areas on which he wrote.

*****

Born in Charlton, south London, on 26 June 1930 and educated at Colfe's Grammar School, Lewisham, MBP always thought of himself as a Kentish man, and retained something of an accent throughout his life. In a foreshadowing of his future scholarly interests in Chaucer, the pilgrim route to Canterbury passed the top of the road on which his primary school was found, and the *Canterbury Tales* were read to the children. During the war, the family were evacuated to Bath, where they were 'bombed out' and returned to London. As a schoolboy, MBP had been strafed on the high street in Eltham and forced to shelter for his life in a shop doorway. In 1944, the family were bombed out again, and evacuated briefly to Petersfield (Hampshire), then Worthing. As a consequence of these wartime experiences, MBP was ever after an ardent pacifist. In 1941, he had won a scholarship to Colfe's, but during hostilities attended the South East London Emergency Grammar School (which moved around

---

[1] Better known as the historian Sir George Prothero FBA.

as it was also bombed out). As a result, his schooling was severely interrupted. MBP reported that he had been taught Latin nouns and adjectives, but did not learn conjugation of verbs until he finally joined Colfe's for the first time in 1945 when it returned to London after the war. He commented, with typical self-deprecation and modesty, that he had coped with being behind in Latin by memorising the set text (then Virgil's *Aeneid* IV, as it very frequently was), and was amazed at how often it had come in useful thereafter. His post-retirement stint as Professor of Latin at Harvard (1998) suggests that this autodidacticism had indeed served him well.

MBP's tertiary education started with a brief stint at Strasbourg University in 1949, where the distinguished French medievalist Ernst Hoepffner helped to stimulate an interest in medieval literature that had first been aroused in him as a sixth-former by reading H. O. Taylor's book *The Medieval Mind* (first published 1911). This was followed by a spell as a supply teacher in the South East (Greenwich) Division of the old London County Council Education Service, and then undergraduate study at Hertford College, Oxford, where he read for the Honour School of English Language and Literature, graduating with a BA in 1953. Reading the main school of English, rather than the more focused course in medieval philology, gave MBP a broad and sympathetic engagement with the study of literature that suffused his later palaeographical and codicological research, and which laid the foundations for his eminence as a cultural historian. In some autobiographical notes written in the 1990s, and transcribed by Pamela Robinson, he describes his own undergraduate career as the seedbed for his later interests and for the formidably idiolectal set of scholarly tools that he accumulated and wielded with such deftness:

> At Oxford against the advice of my language tutor, I read the modern literature course from Beowulf to 1832. NRK[er] also maintained that this was a mistake but as time went on I ceased to regret it because it made *Pause and Effect* possible. Taught by Dennis Horgan—a clear thinker. My literature tutor F. W. Bateson liked essay on text of RIII—set because we had disagreed over whether a quotation in my essay was accurate or not.

In this account, his two undergraduate tutors represent almost the two hemispheres of MBP's brain: the analytical clarity of Dennis Horgan's thought, and the imaginative range and cultural curiosity of F. W. Bateson melded together. His close Oxford colleague Stephen Wall, who later inherited the editorship of *Essays in Criticism* from Bateson, always praised and valued MBP's literary sensitivity as well as his lexical nimbleness. Wall and he made a formidable, if at first glance unlikely, alliance of tutorial approaches, and taught together with great success for well over forty years, initially at Mansfield College, Oxford, and then as tutorial fellows at Keble.

Bateson obviously saw something in the young Parkes, as the autobiographical notes about his early research career suggest:

> In my third year FWB[ateson] suggested that I do research on something that would bore anybody else, and decided he would talk to NR Ker with whom he recommended I should pursue a BLitt. This brought me into a new world centred on Duke Humfrey inhabited by Neil [Ker], Richard Hunt, Roger Mynors, AB Emden and others: the world of the professional scholar in which research was a craft which embraced both learning and skills (not least of which was writing and thinking both in local as well as broader arguments). It was a world of humanist tolerance as well and I was treated not as a brash young 'candidate', but as an equal (even if I had to be taught, gently, that tact should be employed when criticizing the work of those who were not fortunate enough to have daily access to the facilities of one of the world's finest research libraries) … 1959 finished my thesis and RAB Mynors urged me to enter for Gordon Duff prize which to my surprise I won (managed to mollify my wife by proposing to spend prize money on chairs on principle of having won my laurels I then proposed to rest on them).

'[W]riting and thinking both in local as well as broader arguments' and 'A world of humanist tolerance': the phrases both sound like MBP and also sum up perfectly his pedagogic principles and scholarly methodology. But so does the joke at the end. However seriously he took his work, and however relentlessly he worried away at textual problems and sweated over what he always called his 'deathless prose', there was always a good, if rather donnish, joke to be found somewhere in the process. In the preface to his 1992 monograph *Pause and Effect: an Introduction to the History of Punctuation*, he was childishly delighted that he had come up with the wheeze of thanking John Lennard for his help by adding, in parenthesis '(whose study of parentheses in English printed verse is now available in print)'.[2] The detail was entirely superfluous, but it allowed MBP to stage a cheeky in-joke. During the production of the 1978 Ker Festschrift, which MBP edited with Andrew Watson, I remember his watery-eyed, side-hugging delight when the very erudite essay on the transmission of some Ciceronian texts by Richard and Mary Rouse (whom MBP always referred to as 'Rouse and Spouse') came back from Scolar Press with the working title of 'Rouse and Rouse: the posterior academics'. It did not get changed.

Still only in his early twenties, MBP had married Ann Dodman in 1954 (she predeceased him in 2009), and their two sons Neil and Martin were born in 1955 and 1956 respectively. Perhaps because of his new family responsibilities, after the publication of his 1955 essay for *Medium Aevum* (on manuscript fragments of Wycliffite sermons), he worked for the family export business, gaining 'experience in

---

[2] A full bibliography of MBP's publications up to that date, including all of those mentioned here, is included in the Festschrift prepared for his retirement: P. R. Robinson and R. Zim (eds.), *Of the Making of Books: Medieval Manuscripts, their Scribes and Readers: Essays Presented to M. B. Parkes* (Aldershot, 1997). Some of his later essays are collected in P. R. Robinson and R. Zim (eds.), *Pages from the Past: Medieval Writing Skills and Manuscript Books by M. B. Parkes* (Farnham, 2012).

the commercial world (mainly involving the administration of accounts)' as he put it in his CV. In 1957–8, MBP was asked by C. R. Dodwell (then about to leave Lambeth Palace Library for Cambridge) to look after the Lambeth Archives, which consisted principally of identifying and shelving groups of documents relating to the Faculty Office and Vicar General's Office. He was also an unpaid consultant in the Kent County Archives Office, under the guidance of Felix Hull. During these years, he published short handlists to the records of the Vicar General's Office and Faculty Office at Lambeth, and to the fragments of medieval manuscripts in Maidstone. His BLitt (with a title of distinctly unParkesian verbosity: 'A study of certain kinds of scripts used in England in the late fourteenth and the fifteenth centuries, and the origins of the "Tudor Secretary" hand') was awarded in 1959. The autobiographical notes remark on the humane tolerance of his own graduate training, and are a striking insight both into his own style as a supervisor and into his sense of the hard work of writing up for publication:

> Whilst a first year graduate student Neil Ker made me write on fragments and insisted I publish a piece on Wyclif fragments in *Medium Aevum*. This taught me how to write—met the formidable editor at the time (88 year old C. T. Onions), was relieved that after Neil's stringent comments on my piece Onions made only one alteration. Thoroughly professional training by NRK[er], RWH[unt], FW [Bateson]. With both NRK and RWH [I was] exposed to their formidable learning, but more important I learnt the value of the 'haggle' to refine ideas and achieve a greater insight and precision. It is fun to find out, [but the] work starts when you have it write it up.

Haggling, brooding, gutting: the Parkesian vocabulary of scholarly meditation was full of monastic rumination and reflection, as well of scholastic argument and disputation. He strove for the perfect balance between *scientia* and *sapientia*, even when listening to undergraduate essays. (He never took them in for marking, saying that he would only lose them if he did, but was an astonishingly acute and attentive listener, even if you only ever got to the end of your second paragraph.) As a graduate student, being trained by MBP in the handling and analysis of manuscript books was an enthralling if sometimes rather scary process. His way with books was, in a word currently popular among university administrators, 'robust'. There was a legend that he had once been shown a red card at the British Museum for 'over vigorous collation' of a manuscript. And to watch him open a manuscript to examine the stitching for purposes of collation was to remember that passage in the Gawain poet's *Clannesse* where the destruction of Sodom and Gomorrah is likened to the explosion of a manuscript binding:

> Þe grete barrez of þe abyme he barst vp at onez,
> Þat alle þe regioun torof in riftes ful grete,

& clouen alle in lyttel cloutes þe clyffez aywhere,
As lauce leuez of þe boke þat lepes in twynne. (963–6)

'as leaves fall away from a book that splits in two'

The early 1960s saw MBP as a busy Oxford college tutor at Mansfield and, from 1965, at Keble. His first university-level post, as a lecturer in English, began in 1964, and it was not until 1971 that he became the *ad hominem* Lecturer in Palaeography in the Faculty of English. Having been awarded a DLitt in 1985 (becoming the first serving member of the English faculty since Dame Helen Gardner to achieve this distinction), he was promoted to Reader in 1993 and became the first full Professor of Palaeography in the history of the university in 1996. Throughout these years, he was tirelessly teaching palaeography, codicology and transcription to generations of graduate students not just in English but in most of the major humanities faculties. He further refined the categories to be used in describing a manuscript book that had been developed by his mentor, Neil Ker FBA, and his checklist became a benchmark for many generations of manuscript scholars in their own work and teaching. He kept attendance registers for these lectures from the mid-1960s until his retirement, and they are an international roll call of the subject and of medieval studies more widely. Turning their pages one sees names that have become the standard-bearers for modern book history and codicological scholarship. For not only did MBP instruct and train Oxford graduates, but visiting faculty from all over the world would sit in and audit his classes.

Rising in time for his graduate lectures at noon, he would then lunch in college and teach undergraduates until dinner on high table at Keble (to which he invariably arrived late, his silk BLitt gown increasingly tattered and green with age). Only then would he settle to his own work until the small hours, unless disturbed, as he often was, by students seeking a sociable pipe and glass of single malt whisky late in the evening. (His generosity of spirit was as remarkable as his generosity with spirits was legendary.) Tutorials often ran hours late, interspersed by the arrival of anonymous dinner guests who would be revealed later as, for example, the prefect of the Vatican Library (Leonard Boyle) or a major scholar on the poem about which the students had just been talking. MBP was an exceptional teacher and a brilliant supervisor, endlessly patient—even if some comment or question might elicit an exasperated cry of 'Oh Gaawd' on an outward breath, as he reached for his tin of tobacco to buy himself time while he worked out the best way to untie the intellectual knot with which he had been presented. He had a genuinely inspired capacity to get to the nub of the point, no matter how muddled its articulation, and to give the student or a colleague a reference or nudge in the right direction that would precipitate productive further thought. In the autobiographical notes, MBP comments:

I enjoyed teaching. I am proud to say that I have learned something from every pupil I have taught even from one or two who never made it to the first degree. (The need to comprehend their difficulties in order to explain sharpened my own mind.) I rejoiced in their discoveries and their results. The interesting number of those interested in manuscript studies broadened my own interests, and as they developed their own individual responses to source materials taught me new ways of responding to problems.

It was that essential intellectual humility, that eagerness to explore new ways of seeing an issue, that made him such an effective teacher. He never tried to create ventrilo-quial simulacra of himself. Rather, he had an iron determination to allow the distinct-ive critical voice of the student to come through, to act as an intellectual midwife to the processes of thought with which they were grappling. Working over a draft or discussing the outline of a chapter, one always felt that he was treating you as an equal, that you were part of a shared quest for the truth, that you were both listening for the audible click when an argument finally moved into its definitive shape. To see his eyes light up when he saw the thread of an argument leading off into the unknown was always exhilarating. He inspired trust and confidence in his many pupils, encour-aging them to strive for the best and to tap into and develop their potential, which he had an uncanny ability to divine. His kindness to generations of graduate students, male and female, is reflected not only in the countless acknowledgements of him in published work, but also in the way that so many of them kept in contact with him by phone, or on visits to Oxford, and by the deep affection and concern that many of them showed for him in his declining years.

Through all these years as a college tutor, MBP researched tirelessly and published steadily, and his international reputation grew. For medievalists of many disciplines, and from many countries, no visit to Oxford in these years was complete without an audience with MBP; his gatehouse room over the lodge in Keble buzzed with gossip and serious scholarship, laughter and liquor. Among many international honours and distinctions, he was elected Fellow of the Society of Antiquaries of London in 1971, a Fellow of the Royal Historical Society in 1977, a member of the Comité Internationale de Paléographie Latine in 1986, a member of the Wolfenbütteler Mediävistischer Arbeitskreis (also in 1986), a Corresponding Fellow of the Medieval Academy of America in 1992, a member of Council of the Early English Text Society in 1995 and finally—many felt rather belatedly—a Fellow of the British Academy in 1993, at the age of sixty-three. Much in demand internationally, he was a Visiting Professor at Konstanz (1974 and 1980) and Minnesota (1991), a Visiting Fellow at the Institute for Advanced Study at Princeton (1996) and Professor of Latin at Harvard (1998), and he guest lectured all over the USA, including at Summer Institutes of Palaeography at the Medieval Academy of America, UCLA (1978), University of Pennsylvania (1976) and Harvard (1988). MBP loved to travel, packing up his trusty bright yellow Volvo estate for trips to Konstanz or wherever. A much retold (and no doubt lovingly burnished) story recorded that on one of these trips, having locked himself out of the car somewhere in rural France, he attempted to break in using a wire coat hanger, all the while reassuring the gathering crowd in his Kentish-accented French 'Je suis cambrioleur spécialiste' (I am a master burglar). In the UK, he supported the teaching of palaeography nationally by serving on various subcommittees of the Standing Conference of National and University Libraries, the University of London's Board of Studies in Palaeography (1978–88) and as a Special Lecturer in Palaeography in Durham (1972), London (1976) and Queen's University Belfast (1966–9). At a time when the discipline was under pressure and its survival as a component in research training in question, MBP was a restless advocate for its pedagogic centrality, almost as a reflex of his own growing understanding of the ways in which it made possible a detailed and incisive cultural history of medieval reading founded on secure and rigorous technical roots.

Although his list of publications had been gently accumulating for some time, it was the appearance of *English Cursive Book Hands 1250–1500* (Oxford, 1969; reprinted Scolar Press, 1979) in the Oxford Palaeographical Handbooks series that signalled the arrival of MBP as a leading taxonomist of English hands. Its opening essay is a brilliantly lapidary account of medieval English book history. Perhaps more pragmatic than many of the continental codicologists and palaeographers, MBP's descriptive taxonomy of hands is still the gold standard for describing the scripts found in books produced in later medieval England. I have a theory that you can tell somebody trained in Parkesian methodology if you show them a plate of a mid-fifteenth-century mixed hand: the Parkesians will always describe it as anglicana with secretary features, while others will see secretary with anglicana features. MBP felt that the compression and concision necessary to fit the text of *English Cursive Book Hands* into the available space had had a bad effect on his own written style: 'the necessary process of compressing my thesis into few pages and trying to get the balance of the other scripts right caused a lot of agony, ruined my prose style, took 8 years, and I never fully recovered', he writes in the autobiographical notes.

There are, indeed, few wasted words in MBP writing. He never used a word processor or a computer, and only rarely ventured onto a typewriter. He used to write each paragraph longhand onto a separate sheet of blank white file paper, leaving space at head and foot for any conjunctions, disjunctions or prepositions that might prove to be necessary. Then the sheets could be rearranged into the desired order and, when he had got the sequence of ideas right, he would number the pages/paragraphs and send them off to his near neighbour, the formidable and indefatigable Mrs Templeton, for typing. In fact, there are very few conjunctions and connectives in his published works: the limpid clarity of his writing usually did not need them. The prose is crisp, brisk and possessed of an arrow-straight logic. ('Logic, simplicity and clarity, and the greatest of these is clarity', was a favourite aphorism.) Sentences are short, sinewy and often pithy. In *Pause and Effect*, for instance, the bald statement that 'The forms of worship are the chief memorials and declarations of Christian doctrine', has a grandeur that Cranmer might have envied for the *Book of Common Prayer*, as well as being simultaneously insightful and provocative about the nature of the liturgical branches of Christianity. While the statement that 'rhetorical analysis is concerned with the rhythm and shape of a discourse' is the sort of observation that Quintilian would have been proud of, and encapsulates *in parvo* MBP's interest in the movement of literary language and the microcosmic and macrocosmic shape of texts as imaginative entities and as words on the page.

Social scientists tell us that the horizon of scholarly visibility, that period of time in which a publication stays in view and is used by the scholarly community, has dropped from about fifteen years or more a generation ago to about seven or eight years now. But MBP's publications remain part of the bedrock of the discipline, widely cited and read with profit and delight. Throughout the 1970s, he published a series of hugely influential essays and book chapters, marked by notably economic expression which masked huge depths of research and scholarship and guided by an increasingly well-calibrated gut instinct about how books were made and used. Indeed, his publications amply demonstrate how within medieval studies the field can still be changed or redirected by a seminal article or book chapter rather than by a full-blown monograph. His 1973 essay 'The literacy of the laity', for example, changed the way that pragmatic literacy was talked about in connection with later medieval English lay readers, and probably made possible—and certainly gave power and plausibility to— much of the work on lay readers of religious books that has flourished in the last thirty years. And it did so with remarkable economy: MBP shared with Stephen Wall the ability to let a well-chosen example do the work of a lot of argumentation. In *Pause and Effect*, described by one historian of punctuation as 'probably the single most useful work on the origins of modern punctuation', he wryly observes of manuscript circulation: 'Before the advent of printing, a text left its author and fell among scribes.' His chapter on the layout of verse was a foundational study of the changing *theoria* and *praxis*, and is particularly magisterial in its command of the subject across many languages and centuries. But the whole book resonated and was admired far outside of medieval English studies.

The famous 1976 essay 'The influence of the concepts of *ordinatio* and *compilatio* on the development of the book', published in the outstanding and still invaluable Festschrift for Bodley's long-serving Keeper of Western Manuscripts Richard Hunt FBA, taught several generations of scholars how to look at paratext and *mise en page* and to make meaningful and precise comments about how they contribute to the transmission and reception of texts. His 1978 collaboration with Ian Doyle FBA, on the production of copies of the *Canterbury Tales* and *Confessio Amantis*, published in another superb Festschrift, this time edited by MBP with Andrew G. Watson for his friend and mentor Neil Ker, is a foundational essay for the modern study of both poets, and underpins the recent identification of many of their busy scribes as officers of the London Guildhall and serial copyists of Middle English literature, with Doyle and Parkes's Scribe D now revealed as John Marchaunt, Common Clerk of the City of London in the early fifteenth century. Born of a deep and lifelong mutual respect and affection, their collaborations produced work that regularly squared the intellectual circles, their intellectual and professional strengths and reflexes, instincts and cautions admirably complementing each other. How two such distinguished scholars, whose prose styles could not be more different, came to create such a milestone of modern scholarship in Middle English is probably itself worthy of in-depth study of their working drafts. How many subordinate clauses died on that battlefield will never be told. But we are all much the richer for the battle having been fought, and the outcome remains a classic. Doyle and Parkes's symbiotic working relationship was long lasting: for decades, they regularly consulted on datings and hands, combining their astonishing visual archives to challenge, nuance and fine-tune each other's arguments.

MBP's fascination with the copying and circulation of Chaucer also manifested itself in a series of important printed facsimiles of key manuscripts. He was a central player in the popularity of such facsimiles, from the development of new methods of high-quality photographic reproduction in the late 1970s until widely available digital surrogacy rendered them largely redundant. His analytical work in these volumes made the facsimiles not only invaluable teaching aids, allowing students to learn palaeography and codicology remotely, but also, through regular, detailed and systematic collaboration with other scholars, integrated the discipline into the wider critical activity that was encouraged by the availability of such books. His 1978 palaeographical description and commentary for the facsimile of Cambridge, Corpus Christi College MS 61, which has the famous frontispiece allegedly showing Chaucer reading *Troilus and Criseyde* to the royal court, set new standards for the genre. In 1979, MBP again collaborated with Doyle, this time on the introductory and ancillary materials for a facsimile of the Hengwrt manuscript of the *Canterbury Tales*, and in 1980 with Richard Beadle on a facsimile of Cambridge, University Library MS Gg. 4. 27, another important copy of Chaucer. In 1988, he moved his focus earlier to provide the palaeographical commentary for a facsimile of *The Épinal, Erfurt, Werden and Corpus Glossaries* for the Early English Manuscripts in Facsimile series. Finally, in 1996 he produced with Judith Tschann a facsimile of

Oxford, Bodleian Library MS Digby 86 for the Early English Text Society Supplementary Series.

One of MBP's most noteworthy attributes, certainly when viewed from an age of ever increasing nano-specialisation, was his ability to write authoritatively across many centuries of book production. From his work on the Leiden riddle in the very first number of *Anglo-Saxon England* (1972), through Boethius (1981) and his still highly regarded 1982 Jarrow Lecture 'The Scriptorium of Wearmouth-Jarrow', and on to the Oxford manuscript of *The Chanson de Roland* (1985), as well as the production and dating of the *Ormulum* (1983), MBP ranged widely across languages and centuries. The diastolic and systolic pulse of his scholarship alternated intense and highly professional scrutiny of specific cases with increasingly confident cultural generalisations built upon those minute particulars.

That oscillation was both conscious and deliberate. In an unpublished talk delivered to the Association for Manuscripts and Archives in Research Collections in 1995, MBP commented:

> What is nowadays referred to fashionably (if not always meaningfully) as 'the sociology of the text', whether handwritten or printed, has for many years been regarded at Oxford as the most important element in the intellectual formation of research students. The study of primary sources in their proper contexts introduces a nexus of disciplines, new to many students, each with its own procedures and principles and its own conceptual framework …

> Palaeography is not an objective science, but as with all humanities subjects and subjective study with its own terminology, still in the process of development.

I suspect that the echo of A. E. Housman ('Textual criticism is a science, and since it comprises recension and emendation, it is also an art') is entirely deliberate, as MBP cherished Housman's acerbic comments on the need for judgement: 'the worst of having no judgment is that one never misses it … Because a man is not a born textual critic, he need not therefore act like a born fool.' But MBP made a signal contribution to the terminology of English palaeography, the persistence of which is witness to his wisdom and judgement in the development of his taxonomy.

In the mid-1990s, MBP became fascinated with the history of reading, engaging enthusiastically with its continental origins, and writing powerfully and well in Italian and in English on it and on the lexical range of terms for reading in medieval languages (including Old English) before this became embedded as a ubiquitous feature of the newly defined field of the History of the Book in anglophone scholarship. Very few UK book historians wrote as well on the subject as Parkes did, because few of them had the deep and wide archival resources garnered from the first-hand examination of paratexts and hypertexts that decades of palaeography and codicology had given him, and few had thought as hard about issues of *ordinatio* and *mise en page*.

MBP's interest in the history of reading was no doubt also sustained by the fact that throughout his career he had a fascination with the book provisions of the medieval University of Oxford. Perhaps it was also fuelled by the fact that Hertford, his undergraduate and graduate college, stands in Catte Street, epicentre of the medieval book trade in the city. From his 1961 publication on the itinerant scribe Henry Mere, to work on the aids to scholarship developed by Oxford friars for the 1980 Bodleian exhibition in memory of Richard Hunt FBA, he had been exploring Oxford's book trade. In 1987, he gave the Robert F. Metzdorf Memorial Lecture at the University of Rochester, New York, 'Book Provisions and Libraries at the Medieval University of Oxford', and this fed directly into the magisterially under-stated chapter 'The Provision of Books' in the volume of *The History of the University of Oxford* (Oxford, 1992) covering the late medieval period. Like many of the contributions to that remarkable volume edited by Jeremy Catto and Ralph Evans, the modesty of the argument's articulation belied the huge erudition and range of scholarship and primary research that it contained. One of his secondary school teachers wrote in a 1949 reference that the young MBP 'has been my chief assistant librarian at Colfe's during the past two years and has shown in this office, as indeed he has shown in other school activities, a marked capacity for organizing the work and for controlling those who were serving under him'. The boy who had been school librarian retained a fascination with the institutional provision and care of books throughout his life.

No surprise then that MBP served as Fellow Librarian of Keble for nine years (1965–74). This was at a time when an extension to the library was built in order to make it more accessible to students. In the notes, he comments that 'I shall probably go down in history as the last librarian of the College to move every book with my own hands. In the process I discovered that we had some ninety MSS (instead of the sixteen listed) and realized that a catalogue was needed.' In 1979, therefore, he published *The Medieval Manuscripts of Keble College Oxford*. One of the great monuments of the modern cataloguer's art, this book set new standards in the detail and format of manuscript description and analysis. Since his arrival in Keble in 1965, MBP had been exploring and examining the college's manuscript collection, which was, he discovered, unusually rich and varied for such a young foundation. Over his years as Fellow Librarian he had painstakingly sifted the books (often in his college study, occasionally propping open the door to let out the pipe smoke) and evolved a matrix for his descriptions that was elegant as well as exhaustive. Of course MBP had his favourites, notably the Regensburg Lectionary (Oxford, Keble College MS 49) made in around 1267–76 for a convent of Dominican nuns, and which Keble has recently digitised in his memory. At his funeral, one of the college's Books of Hours was laid open on his coffin by the Warden, in recognition of his outstanding service to the collection. The balance between economy and detail that he struck in the Keble catalogue entries makes the book an absolute pleasure to use, and, as with all great catalogues, a mine of arcane and pertinent information on all sorts of topics. It is a reflection of and a tribute to an extraordinary ability to see the big picture while being in total command of small details.

Like many of MBP's projects, from the Ker Festschrift through to *Pause and Effect*, the Keble catalogue was published by Scolar Press and supervised by Sean Magee. Magee produced a spoof appendix for what was universally known as the Kerschrift, with the running head 'SONNENSCHEIN The "Anterior" Academics', playing on the Rouses' contribution and MBP's invariable telephone greeting of "ello sunshine'. It is full of in-jokes about a wayward pair of scribes, P (MBP, of course) and W (Andrew Watson). Describing P, the note says 'His hand is a bastard, partially cursive *anglicana deformata*'—the joke very appropriately coming in part through the punctuation, parodying MBP's fastidious use of commas. Scribe P is said to be identified in one book as 'scriptor in tunicae rubissimae qui loquitur per diem et noctem', referring to MBP's penchant for brightly coloured shirts and lengthy conversations. The playful and highly effective partnership between Magee and Parkes resulted in books that were unusually pleasing to use as well as exceptionally scholarly in their content, and Magee reflected on this relationship in his contribution to the affectionate portmanteau introduction to the Festschrift for MBP edited by Pamela Robinson and Rivkah Zim for his retirement in 1997 under the title *Of the Making of Books: Medieval Manuscripts, their Scribes and Readers*, and published of course by Scolar Press. This book stands alongside the Festschriften for Richard Hunt FBA and Neil Ker FBA as an essential volume for those interested in medieval book history, and gave MBP's many pupils and friends a chance to pay concrete tribute to his support, leadership and scholarly example.

Pause and Effect's interest in the palaeographical 'grammar of legibility', explored further in essays and lectures in those years, flows into MBP's last book, *Their Hands before our Eyes: a Closer Look at Scribes* (Aldershot, 2008), where it is augmented by an advanced fascination with distinctive (and he felt inimitable) scribal spacings between letters. Based on the J. P. R. Lyell Lectures he had delivered in the University of Oxford in 1999, it was much augmented, enhanced and worried over before its appearance. In retrospect, it is possible to see the beginnings of the vascular dementia that eventually took his life in the travails he underwent to finish this book, and when it came to proofing and indexing it he needed help from Zim and Robinson. But it is full of observations that deserve and reward Parkesian brooding over, a store of forensic wisdom and synthesised first-hand experience of thousands of medieval books, deployed using precise terminology and progressing rigorously through *catenae* of carefully differentiated concepts:

> A scribe's interpretation is most evident in the equilibrium of his or her handwriting that contributes substantially to its aspect (the impression made by the handwriting at first sight). Aspect is difficult to analyse, but is created by a scribe's own way of combining different features of the handwriting to achieve a balance between style and function. The extent to which a scribe achieves this equilibrium depends on his own particular talent: on the competence of his penmanship to express graphic ideas in his own way, and on his capacity to respond to the demands of a prevailing sense of decorum.[3]

Late Parkes is like late Beethoven: a thing of spare and austere, almost ethereal beauty. The style is even more pared back, the discussion even more abstract, with its near mystical emphasis on ducts and aspects, its fascination with a scribe's ability to create chiaroscuro effects on the page. Yet the range of examples, the lifetime's expertise in choosing the perfect plate, the eye for the telling palaeographical detail are all still on full and peerless display. This was the work of a man who had long since ceased to have anything to prove, but was looking back at a long career of working with books ('never refuse the opportunity to look at a manuscript', he used to say), and was able to draw a line of gossamer thinness and silken elegance between the particular instance and the general circumstance. It is not an easy book, but it is the late and last work of a master. In a letter to Tessa Webber FBA in 2000, MBP described the Lyell Lectures given the previous year as

> in one sense my 'palaeographical testament'… In particular I want to put a coherent terminology into place. My elders and betters Alan Bishop, Richard Hunt and Neil Ker all thought that my contribution to the subject would be a terminology which encouraged close analysis…. I feel these friendly ghosts looking over my shoulder, but they can be inhibiting sometimes.

A couple of articles on the Oxford book trade and the work of Thomas Hunt were found among his papers and were published posthumously in *The Library* in 2016, but this was his last major publication. The later collection of essays put together by Robinson and Zim has the peculiarly plangent title of *Pages from the Past*, for by the time it appeared in 2012 MBP was already beyond new work. Yet, like the work of the great nineteenth- and twentieth-century medievalists whom he so much revered and respected, MBP's body of writing will stand scrutiny for generations to come.

In a sermon preached in Keble College Chapel on St Mark's Day 1879 to mark the first anniversary of opening of the College Library, Edward King, at the time Regius Professor of Pastoral Theology at Oxford, Canon of Christ Church and later Bishop of Lincoln, had said:

---

[3] Malcolm Parkes, *Their Hands before our Eyes: a Closer Look at Scribes* (Aldershot, 2008), p. 87.

We need now men who can read and copy MSS … to carry out any research-work, in the way of criticism and amended texts, we need the help of those who have time and skill to examine unprinted matter … in the interests of literature, and at the disposal of the society. That such a student should have been found in Keble College, and in connection with this Keble Library, would, I venture to think, add another ground of hope for this hopeful society.

MBP quoted this sermon in the introduction to his Keble catalogue. In an unpublished note on the history of palaeography teaching at Oxford, he had observed that King's hope had been fulfilled in the 1930s by the Keble men Noel Denholm-Young and T. A. M. Bishop, who had taught palaeography and diplomatic studies at Oxford and Cambridge respectively. But Malcom Parkes, mindful and protective of Keble's long tradition of medieval scholarship, was himself the truest embodiment of King's aspiration, able to bridge and synthesise the disciplines and skills of literature, textual criticism, language and palaeography into a single hermeneutic tool of unusual power and grace. He is much missed, as an outstanding scholar, teacher and friend. Each chapter of *Their Hands before our Eyes* concludes with a medieval Latin manuscript colophon. At the end of the last chapter, he offered a sort of palaeographical epitaph, rounding off a career of energetic service to his college and university, and his striking scholarly contribution to medieval studies:

Iste libellus scriptus est per fidelem scriptorem in universitate Oxonienesis.

Acknowledgements

In preparing this memoir, I acknowledge the invaluable assistance of MBP's executors, Pamela Robinson and Rivkah Zim, for access to unpublished materials from his papers, and offer thanks to Richard Beadle, David Ganz, Ian Johnson, Alastair Minnis, David Rundle, Tessa Webber and many other colleagues who contributed reminiscences and copies of obituaries.

Note on the author: Vincent Gillespie is J. R. R. Tolkien Professor of English at the University of Oxford. He was elected a Fellow of the British Academy in 2013.

Martin Litchfield West

23 September 1937 – 13 July 2015

elected Fellow of the British Academy 1973

by

ROBERT FOWLER

Fellow of the Academy

Biographical Memoirs of Fellows of the British Academy, XVII, 89–120
Posted 27 September 2018. © British Academy 2018

MARTIN WEST

Martin West was one of the greatest Hellenists of modern times. Although working largely within the boundaries of traditional classical philology—editing texts, writing commentaries, reconstructing literary history—his enormous oeuvre can be called revolutionary in the sense that nearly everything he wrote decisively affected the course of scholarship. Revolutionary without question is his work on the ancient Near Eastern background of Greek literature, culminating in *The East Face of Helicon* (Oxford, 1997). The conviction, evident already in the commentary on Hesiod's *Theogony* (Oxford, 1966), that early Greek literature must be understood in this wider, eastern Mediterranean context, had few advocates before him, mere voices in the wilderness; as a result of his labours, awareness of the Near Eastern background has become obligatory for all Greek scholars. A similar view of Greek myth, philosophy and religion was championed by Walter Burkert (1931–2015), whom West first met while studying at Erlangen in summer term 1960 under the great Hellenist and historian of ancient religion Reinhold Merkelbach (1918–2006). This was a serendipitous meeting, for apart from their lifelong friendship Burkert was the scholar West admired most among his contemporaries. They had in common a determination both to document the formative influence of adjacent cultures upon the Greek, and to trace its ancestry—in Burkert's case, back to early hunting societies and their rituals of sacrifice; in West's case, to the Indo-Europeans, leading ultimately to *Indo-European Poetry and Myth* (Oxford, 2007). Though for this branch of study West could claim many fine predecessors, here too his research was transformative. Overall, the output is astonishing in both quantity and quality; many of these books in themselves would be a satisfactory life's work.[1] No Hellenist is more frequently cited, and West will continue to be cited for as long as the subject survives.

[1] Thirty-five books and three volumes of *Hellenica* (Oxford, 2011–13; they contain ninety-four shorter writings selected from some 550). The book count excludes second editions, partial reprints, translations into other languages and the edited papers of W. S. Barrett, but includes the posthumous edition of the *Odyssey* (Berlin, 2017) which S. R. West has seen through to publication. It includes also *The All Souls Mallard: Song, Procession, and Legend* (All Souls College, 2001), slight though it is, since West counted it as a book in his list of publications. The list (up to 2014) is most easily consulted on the All Souls website, https://www.asc.ox.ac.uk/asc-person-section-publication/79/4/999 (accessed 27 April 2018); there are some differences in the years post-2003 from the list published in P. J. Finglass, C. Collard and N. J. Richardson (eds.), *Hesperos: Studies in Ancient Greek Poetry Presented to M. L. West on his Seventieth Birthday* (Oxford, 2007), pp. xxix–lvi. Note also the lists by category at the end of each volume of *Hellenica* (Oxford, 2011–13). Addenda: 'The classical world', in M. Stausberg and Y. Sohrab-Dinshaw Vevaina (eds.), *The Wiley Blackwell Companion to Zoroastrianism* (Malden, MA, 2015), pp. 437–50; 'Epic, lyric, and lyric epic', in P. J. Finglass and A. Kelly (eds.), *Stesichorus in Context* (Cambridge, 2015), pp. 63–80; 'Early poetry in Cyprus', in G. A. Xenis (ed.), *Literature, Scholarship, Philosophy and History: Classical Studies in Memory of Ioannis Taifacos* (Stuttgart, 2015), pp. 25–36; 'Mythological and political interpolations in Homer', *Eikasmos*, 26 (2015), 13–25; 'The formation of the epic cycle', in M. Fantuzzi and C. Tsagalis (eds.), *The Greek Epic Cycle and its Ancient Reception: a Companion* (Cambridge, 2015), pp. 96–107; 'Odysseus re-routed', *Eikasmos*, 27 (2016), 11–23; 'So what is it to be?', in D. Gunkel, J. Katz,

*****

Martin Litchfield West (Litchfield was the maiden name of his paternal grandmother) was born on 23 September 1937 at Eltham General Hospital; his parents lived at the time in Orpington (then in Kent, now part of Greater London), but moved shortly before the outbreak of war in 1939 to Hampton, Middlesex, where his father, Maurice, was appointed resident engineer at the waterworks operated there by the Metropolitan Water Board. His father's family were from the Home Counties, but that of his mother, Catherine Baker Stainthorpe, were from Yorkshire and Durham. His paternal grandfather, Robert West, lectured in electrical engineering; his maternal grandfather, John Stainthorpe, was a railwayman from Pickering. In a sparkling and typically amusing (unpublished) memoir of his childhood written for his family, West tells how on both parents' sides he could lay dubious claim to royal ancestry: John Stainthorpe's mother was said to have been an illegitimate daughter of George, the second Marquis of Normanby, and thus descended from James II; Robert West traced his lineage (via a great-great-grandmother who had eloped with the gardener) to Sir William Courtenay, '*de jure* eighth Earl of Devon and second Viscount of Powderham Castle'. Consequently, West calculated, he was tenth cousin to the Queen. More than that, the ultimate ancestor of all the royals, Egbert of Wessex, was separated by a mere twenty generations from Woden. 'Once', he confesses, 'in filling up a form for some French biographical reference work, I amused myself by naming him as my ancestor. And so it appeared in my entry: "Ascendance: Woden, dieu germanique".' Among other earthly connections, the mother of the aforementioned Sir William Courtenay was a distant descendant of the brother of Archbishop Henry Chichele, founder of All Souls College, so that when West arrived there he could lay 'tenuous claim to membership of that blessed fraternity, Founder's Kin'. All of this was, he admitted, 'sentimental nonsense', but it does relate to his strong sense of rootedness in his native country, embracing both Germanic and Celtic heritage:

B. Vine and M. Weiss (eds.), *Sahasram Ati Srajas: Indo-Iranian and Indo-European Studies in Honor of Stephanie W. Jamison* (Ann Arbor, MI, 2016), pp. 486–98; 'Editing the *Odyssey*', in C. Tsagalis and A. Markantonatos (eds.), *The Winnowing Oar—New Perspectives in Homeric Studies: Studies in Honor of Antonios Rengakos* (Berlin and Boston, 2017), pp. 13–28; 'Aristophanes of Byzantium's text of Homer', *Classical Philology*, 112 (2017), 20–44; the edition of the *Odyssey* (Oxford, 2017); '*Gilgāmeš* and Homer: the missing link?', in B. Dignas and L. Audley-Miller (eds.), *Wandering Myths: Transcultural Uses of Myth in the Ancient World* (Berlin, 2018), pp. 259–75; (with L. Bendall), 'Evidence from the written sources', in I. S. Lemos and A. Kotsonas (eds.), *A Companion to the Archaeology of Early Greece and the Mediterranean* (Malden, MA, forthcoming); 'What next?', in C. A. Stray, M. J. Clarke and J. T. Katz (eds.), *Liddell and Scott: the History, Methodology and Languages of the World's Leading Lexicon of Ancient Greek* (Oxford, forthcoming).

One does have—at least I have—a deep-seated desire for a tribal identity. I always thought of myself as an Englishman, and now I have learned how I can think of myself as a Briton too. Having grown up during the Second World War, I could hardly avoid being imbued with patriotic feeling, which I have never disavowed. I cannot imagine living permanently in another country, and have more than once turned down invitations to do so.

At the age of four West began his education at Denmead, a private preparatory school about a mile from the family home. While nothing of academic significance seems to have occurred there (the Latin teacher was 'an ignorant thug who knew only the declension of *mensa* and the present indicative active of *amo*'), West's promise must have been obvious, and when he turned eleven one of his teachers persuaded his parents that he should put in for a scholarship at Colet Court, the junior school for St Paul's. Heroic cramming with the aid of a Latin grammar purchased by his father did not compensate for the shortcomings of the thug's instruction, so West did not win the scholarship; but he was offered a fee-paying place, and his parents resolved to make the necessary sacrifices.

At Colet Court, as at St Paul's, Latin and Greek held pride of place in the curriculum. Here West discovered his interest in languages, including Esperanto, though this proved to be 'the least useful of all the languages I have ever studied'. At age fourteen he invented a competitor dubbed Unilingua, complete with dictionary and a sampling of texts. The memoir also records passions for astronomy, stamp-collecting, coin-collecting, bus-spotting, plane-spotting (passing the wartime test at age five): if the Child is the father of the Man, such matters are worth mentioning here, since one of the outstanding characteristics of West the scholar was his love of complicated formal systems and his ability to impose order on large masses of unruly data. Subjects such as Greek metre, music, manuscripts, mythical genealogies or comparative philology were almost bound to attract his attention. In the case of astronomy, the interest continued throughout his life; his first three professional publications were on topics in historical astronomy, and his expertise informed many writings at all periods of his career, especially the commentaries on Hesiod. Several numbers of a journal *Starry Nights*, which he began at Denmead and continued at Colet Court, survive; partly printed with a hand press owned by West but mostly written in the distinctive, boyish hand that never changed throughout his life, they inform the reader (for instance) how to correct variable star observations for atmospheric absorption, or that on 3 October 1951 Jupiter will be in its most favourable opposition for years, magnitude −2.5. In addition to such instruction, they offer a menu of news, poems, quizzes, art, notes from the editor—everything a proper journal should have. Touches of sly Westian humour appear frequently. Also surviving is an essay, 'A theory concerning the history of the solar system', whose grandiose title anticipates other, equally ambitious ones, such as 'Greek Poetry 2000–700 BC'.[2]

In summer 1951, West sat for and won a scholarship to the main school. As at Colet Court, Latin and Greek dominated the curriculum, with a strong emphasis on composition; a not very balanced education, West allowed, 'but for a budding classicist it was superb'. He was taught by two great teachers, E. P. C. Cotter and W. W. Cruickshank; for the latter West and other ex-pupils composed a Festschrift, a rare honour for a schoolmaster.[3] He raced so far ahead of the other boys (excelling also at mathematics) that he was advanced to the Upper Eighth (as the nomenclature at St Paul's had it), and so sat the four-day examination for a scholarship to Balliol a year early. His teachers thought he might manage to win the lesser award of an exhibition, and his own expectations were not high, but instead of an exhibition he carried away the top scholarship. Kenneth Dover, later President of the British Academy, was then at Balliol and Tutor for Admissions. At the bottom of the formal, typed notification he wrote: 'When I won the same scholarship in 1938 Cyril Bailey wrote to me "Paulinus Paulino tibi gratulor" ["congratulations from one Pauline to another"]. This is the first time I have been able to use the same words as Classics tutor!' Dover departed for St Andrews that summer, so West was never among his pupils at Oxford. West's principal tutors were Donald Russell and Michael Stokes in Greek, Gordon Williams in Latin, Russell Meiggs in Ancient History and Dick Hare in Philosophy. Among his peers were Anthony Leggett, who took a first in Physics the year after he took a first in Greats, and went on to win a Nobel Prize, and Peter Gregson (1936–2015), who became Permanent Secretary first of the Department of Energy, then of Trade and Industry.

*****

The memoir ends with arrival at Balliol, its author describing himself as a 'serious-minded but light-hearted youth of seventeen … already marked with the lineaments of the man I was to be'. The description is apt. Serious about scholarship, of course, he worked hard and without interruption until the last day of his life, taking few holidays. The lightness of heart, however, might not have been obvious to those who knew him only casually. Notoriously taciturn in conversation, he could have been

[2] M. L. West, 'Greek poetry 2000–700 BC', *Classical Quarterly*, 23 (1973), 179–92; repr. with suppl. notes *Hellenica* 1, pp. 1–21. (No Greek poetry actually survives from the period in question.) I owe sight of these documents to Alan Cameron FBA (1938–2017), a keen fellow member of the 'Herschelian Society', as it was called; Cameron and the distinguished Roman historian John North started studying Greek with West at Colet Court on the same day in 1949.
[3] *Apodosis: Essays Presented to Dr. W. W. Cruickshank to Mark his Eightieth Birthday* (St Paul's School, 1992).

mistaken for a typically shy, unworldly scholar, wholly engrossed in his own mental universe. All Souls College, most ivory of towers, elected him Senior Research Fellow in 1991, offering an escape—as West saw it—from the bureaucratic horrors of university life.[4] Here he was truly at home, and the College never had a more devoted Fellow. Yet he was not obsessive like his scholarly hero Ulrich von Wilamowitz-Moellendorff (1848–1931), who worked two hours before breakfast every day. Scholars who take themselves too seriously do not publish spoofs of German philology, in German, in German journals.[5] Scholarship gave meaning to life, but it was not all-consuming for West; he made time for other pursuits (music, literature, cricket). His writings are full of wit and sharp observations of life—not the work of someone disconnected from ordinary human discourse.[6] A secret extrovert seemed to lurk within, who might show himself, for instance, in an *alta voce* declamation of an ancient Greek hymn while touring Delphi;[7] or in theatrical openings to lectures on Homer, in which he would adapt famous Homeric similes to describe the swarm of undergraduates before him (or perhaps regale them with the story of Goldilocks in the style of the Homeric bards); or in mimicry of famous scholars and politicians; or in the enthusiastic discharge of his duties as Lord Mallard at All Souls College (being paraded about, shoulder-high in a sedan chair and singing the Mallard Song, among other things).[8]

West's precocious ability, particularly in languages, was already abundantly clear upon arrival at Oxford. In 1957, he netted a haul of undergraduate prizes (Chancellor's for Latin Prose and Verse, Hertford, de Paravicini, Ireland), though when it came to sitting Greats he took only a second. (At that time, Greats still consisted of philosophy and ancient history; West liked philosophy and found Greek history tolerable, but he was allergic to Roman history. He was allowed to offer a special paper on Homer, for which he received tutorials from E. R. Dodds and C. M. Bowra—a good foundation for postgraduate work, but the First still proved elusive.) Other lifelong pursuits already in place were writing poetry, short stories and music (though he was secretive and somewhat diffident about the latter; a movement from a piano sonata in late Romantic style—his seventh, no less—was performed at his All Souls memorial event, to the complete surprise of most of those in attendance). What was lacking at school, and came with university, was exposure to professional classical scholarship. This

[4] M. L. West, *The East Face of Helicon* (Oxford, 1997), p. xii, quoting two verses from the so-called Orphic tablets in which the initiate exults in having escaped the 'grim circle of deep grief' and won eternal bliss. The book is dedicated to the College, where his fellowship gave him time to acquire the necessary Near Eastern languages.

[5] See *Hellenica* 3, p. 498. To one he signed his name 'M. L. W. Eggheider'.

[6] He assembled his own favourite remarks on scholars' behaviour in '*Obiter Dicta*', *Hellenica* 3, pp. 485–9.

[7] The incident is recounted in the preface to M. L. West, *Ancient Greek Music* (Oxford, 1992).

[8] See West, *The All Souls Mallard*.

took the form of Eduard Fraenkel's renowned seminars, which were a decisive influence. Over forty years later, West recalled:

> Here we saw German philology in action; we felt it reverberate through us as Fraenkel patrolled the room behind our chairs, discoursing in forceful accents. As he spoke of his old teachers and past colleagues—Leo and Norden, Wilamowitz and Wackernagel—it was like an *apparition de l'Église eternelle*. We knew, and could not doubt, that this was what Classical Scholarship was, and that it was for us to learn to carry it on.[9]

Fraenkel's monumental edition of the *Agamemnon* had appeared in 1950; West dedicated his own edition of Aeschylus to his memory, citing in the preface the words Orestes addresses to his dead father in the *Choephori*.[10] The first and most abiding lesson of the seminars was that textual criticism was the foundation of scholarship, a point West emphasises at length in the sequel to the passage just quoted. It was a craft to which he was in any case naturally predisposed, and in which he showed himself a master from his first book (written at age twenty-nine), the edition of Hesiod's *Theogony*. Of his thirty-five books, eighteen are editions or commentaries, to which one may add his manual of textual criticism and his work on the *Supplement* to the Greek lexicon of Liddell, Scott and Jones.[11] We have the entire corpus of archaic Greek poetry, with the exception of the lyric poets (i.e. Sappho, Alcaeus and the authors edited in D. L. Page's *Poetae Melici Graeci*: Oxford, 1962) in authoritative editions by West, much of what he edited accompanied by commentary, and much of it he translated too, including the lyric poets (who also received invaluable attention in numerous articles; had he lived, his next project was to be an edition).

As is clear from the above quotation, it was in these seminars that West learned fully to value the achievements of German classical scholarship. Again, it was a case of a call falling upon willing ears. His instinct was always to find concrete answers to concrete questions. He had little sympathy with literary or anthropological theory evident in much French and American work throughout his career, and he scorned

---

[9] M. L. West, 'Forward into the past', in Finglass, Collard and Richardson, *Hesperos*, pp. xx–xxviii at xxi. The address is wrongly identified as the acceptance speech for the International Balzan Prize for Classical Antiquity, 2000; it was composed on that occasion, but the actual acceptance speech was different. Both are available at http://www.balzan.org/en/prizewinners/martin-litchfield-west (accessed 14 November 2017). On Fraenkel and his seminars, see S. West, 'Eduard Fraenkel recalled', in C. Stray (ed.), *Oxford Classics: Teaching and Learning 1800–2000* (London, 2007), pp. 203–18.

[10] M. L. West, *Aeschyli Tragoediae cum incerti poetae Prometheo* (Stuttgart, 1990), dedication, p. lv, quoting *Cho.* 315–19 ('What might I say or do to bring you here on a gentle wind, from far away where your place of rest detains you?').

[11] M. L. West, *Textual Criticism and Editorial Technique Applicable to Greek and Latin Texts* (Stuttgart, 1973); acting editors, H. G. Liddell. R. Scott, and H. S. Jones, *Greek–English Lexicon: a Supplement*, ed. E. A. Barber, with the assistance of P. Maas, M. Scheller and M. L. West (Oxford, 1968). On the latter (and its 1996 successor), see West, 'What next?'.

those who thought 'positivism' a dirty word. His own work combined the best of the older English and German traditions: the former characterised by textual emendation based upon the most precise knowledge of the classical languages; the latter founded on the conviction that to explicate a text, one must bring to bear any and all data that may shed light upon its problems—literary texts of any genre, inscriptions, papyri, works of art; linguistics, historical context, philosophical background and so on. This tradition reached its apogee in the career of Ulrich von Wilamowitz-Moellendorff, whose portrait hung in West's study and with whom he has often been compared for learning, impact and sheer philological power. He shared with Wilamowitz also a lack of pedantry, a sharp focus on primary sources and a talent for bold hypothesising. West did not, however, embrace the ideal of German *Altertumswissenschaft*, which (naively, as it must seem nowadays) sought to know the whole of antiquity in order to understand any of its parts. In theory, in this scheme all subdisciplines (literature, history, philosophy, art) were on equal footing, supporting a *Gesamtbild* of antiquity built up in the all-conquering scholar's mind. West was a literary scholar first and last. All of his books were on literary or closely related topics. His brand of literary history, to be sure, made room for mythology (wellspring of Greek literature) and early philosophy (whose cosmology was a calque on Greek myth); but he wrote nothing on political history or art, for instance, and his interest in mythology did not induce him to write on the cultic aspects of ancient religion (he was himself antipathetic to modern religion). West also described himself simply as a scholar of early Greek poetry. He was not much interested in the artificialities of the post-classical poets and wrote little on prose authors (and almost nothing on Latin). Of course, he knew the later texts well and published notes of various kinds throughout his career; as a parergon he edited the *Anacreontea*, a Roman and early Byzantine collection of poems notionally in the style of and attributed to the archaic poet Anacreon.[12] But the focus remained relentlessly on early Greece. His only book with substantial post-classical content, *The Orphic Poems* (Oxford, 1983), had as its ultimate purpose the explication of the Orphic texts and traditions of the early period, for which it was necessary to unravel the tangled skeins of their later testimonies and echoes (see also below p. 112).

Fraenkel's seminars must also have confirmed West in his inclination to pursue research. After graduating he embarked on a DPhil, at a time (1959) when doctoral students were quite rare birds. His supervisor was Sir Hugh Lloyd-Jones FBA (1922–2009); West was Lloyd-Jones's first supervisee and the only one before he was appointed to the Regius Chair of Greek in Oxford (1960). West was also the first holder of the Woodhouse Junior Research Fellowship at St John's College, Oxford. The relationship

[12] M. L. West, *Carmina Anacreontea* (Leipzig, 1984; 2nd edn 1993); see also 'The *Anacreontea*', *Hellenica* 2, pp. 385–90.

with Lloyd-Jones as supervisor was highly satisfactory, as West records in a lecture delivered in his memory; he also wrote the *Times* obituary (9 October 2009).[13] West later disagreed with Lloyd-Jones about several issues close to the latter's heart: the authenticity of *Prometheus Bound*, the conception of Zeus in Aeschylus and the idea of the inherited curse in Greek literature (the second of these being the subject of the lecture just mentioned, and the third being published, somewhat surprisingly, in Lloyd-Jones's Festschrift).[14] The review of the Oxford Classical Text of Sophocles produced by Lloyd-Jones and Nigel Wilson, while stressing more than once the great superiority of the edition over others, did not pull its considerable critical punches.[15] These exchanges put a strain on the relationship, but West's obituary of Lloyd-Jones was a warm and sincere tribute; he had fond words to say also in his Balzan acceptance speech.[16]

As a doctoral student West needed little supervision, and the commentary on the *Theogony*, which became the book of 1966, easily won the Conington Prize in 1965 for the best classical dissertation of the year. It was worth half a dozen doctorates. The poem, by a late eighth-century BC (on West's dating) Boeotian poet, recounts in some 1,000 hexameters the birth and genealogies of all the gods; it is both a theogony and a mythological cosmogony accounting for the origins of the universe. West's commentary opens with a survey of worldwide theogonic poetry; a section exploring the links with Anatolian and Mesopotamian literature foreshadows *Early Greek Philosophy and the Orient* (Oxford, 1971) and especially *The East Face of Helicon: West Asiatic Elements in Greek Poetry and Myth* (Oxford, 1997). (In the Balzan acceptance speech, West credits Merkelbach with opening his eyes to this vast new panorama.) He expresses here his conviction that Hesiod antedated Homer, a view he defended stoutly throughout his life.[17] The treatment of the manuscripts, papyri,

---

[13] *Hellenica* 2, pp. 175–7, and 3, pp. 482–4.

[14] 'Ancestral curses', reprinted in *Hellenica* 3, pp. 287–301 (see also *Hellenica* 2, pp. 164–70). Lloyd-Jones responded in 'Curses and divine anger in early Greek epic: the Pisander Scholion', *Classical Quarterly* n.s. 52 (2002), 1–14, reprinted in *The Further Academic Papers of Sir Hugh Lloyd-Jones* (Oxford, 2005), pp. 18–35.

[15] M. L. West, 'The new Oct of Sophocles', *Classical Review*, n.s. 41 (1991), 299–301. West reviewed many books throughout his career; he always balanced praise with blame where merited, but his criticism could be expressed in cutting, sometimes insulting terms. Lloyd-Jones and Wilson responded in *Sophocles: Second Thoughts* (Göttingen, 1997), accepting some of the points made, other times robustly defending their position. See also Lloyd-Jones's review of West's *Aeschylus* in *Gnomon*, 65 (1993), 1–11, reprinted in *The Further Academic Papers of Sir Hugh Lloyd-Jones*, pp. 163–80.

[16] Above, n. 9.

[17] West, *Hesiod, Theogony*, p. 46, with n. 2. At the other end of his career see 'Echoes of Hesiod and elegy in the *Iliad*', first published in *Hellenica* 1 (i.e. 2011), pp. 209–32 (p. 232: 'Those not yet accustomed to the idea of the Homeric epics as products of the seventh century should shake themselves down and come to terms with it'). Many scholars still accept the traditional order.

indirect traditions, metre and dialect are models of their kind. The judgement and learning displayed in the constitution of the text and the commentary, which is a treasure house of information and insight, are mature far beyond their author's years. There is room for disagreement of course, particularly on the amount of interpolation (West diagnosed much less than was fashionable), but the most qualified reviewer, a scholar many years West's senior, while having quite different views, was unstinting in his praise.[18]

Yet this was not West's first publication; he started in 1960 with an article on Anaxagoras. By 1964 he had written on Lucretius, Persius, the Orphic hymns, Empedocles, Musaeus, Nonnus, Hesiod (notes, and an article on the manuscript tradition), Hecataeus, Apollonius of Rhodes, Oppian, Nicander, Quintus of Smyrna, Archilochus, Megasthenes and Servius, along with a substantial article on Presocratic cosmologies and two lexicographical notes. Three of the articles were in German. In 1964, there appeared in volume 11 of *Greece & Rome* (pp. 185–7), the now legendary tour de force 'Two versions of Jabberwocky', one in Homeric hexameters, one in Nonnian ('taking due account of the Humpty-Dumpty scholia'). In 1965, he published substantial articles on Trypho, Alcman, the Dictaean Hymn to the Kouros and (with Merkelbach) the Hesiodic Wedding of Ceyx, plus a note on Euripides; the year the *Theogony* appeared, he also published 'Conjectures on 46 [*sic*] Greek poets', plus other notes and the usual clutch of reviews.

Thus began the publishing career that saw its author elected Fellow of the British Academy in 1973, still, at the time of this writing, the second-youngest person to be elected after Bernard Grenfell (1905, aet. 35). In 1963, West was elected Fellow and Praelector in Classics at University College, Oxford; in 1974, he was appointed to the Chair of Greek at Bedford College, London, and later at the merged Royal Holloway and Bedford New College. He moved to All Souls in 1991, retiring formally in 2004 but remaining an active presence in the College to the end.

In what follows, I will assess West's contribution according to the same categories and in the same order as he arranged his *Hellenica*, before concluding with remarks on the general character of his scholarship.

*****

First, then, epic. West left us editions and commentaries on Hesiod, *Theogony* and *Works and Days*; with Merkelbach, an edition of the fragments of other poems by, or attributed to, Hesiod;[19] a book on the fragments of the Hesiodic *Catalogue of Women*;

---

[18] F. Solmsen, *Gnomon*, 40 (1968), 321–9.

[19] Only a small portion of classical literature survives intact, through the medieval manuscript tradition; for lost works, we are dependent on fragments, i.e. partial quotations in surviving authors, or scraps of ancient copies on papyrus or occasionally other materials. It is a basic task of classical philology to assemble the scattered fragments of lost works and attempt their reconstruction.

two Loeb editions (one of the fragments of archaic epic other than the *Iliad* and *Odyssey*, and another of the *Homeric Hymns*, ancient lives of Homer and Homeric apocrypha); a commentary on the epics about Troy in the first Loeb volume; editions of the *Iliad* and the *Odyssey*; a book on the textual transmission of the *Iliad*, with a commentary on hundreds of selected passages; a book on the composition of the *Iliad*, and another on the composition of the *Odyssey* (each containing many more comments on individual passages); plus numerous articles.[20]

It is convenient here to speak in detail of West's Homeric studies. The problem confronting the editor is not, as often with Greek and Latin authors, a paucity of evidence, but rather an abundance of it: over 1,500 fragments of ancient copies of the *Iliad* and 566 of the *Odyssey*, as well as hundreds of fragments of other texts quoting Homer;[21] hundreds of medieval manuscripts; a forest of ancient and medieval commentary; a mountain of modern scholarship. Traditionally, the task of the editor is to reconstruct as closely as possible the text as written by its author; in the case of Homer, because of the texts' background in oral poetry and the fluidity of ancient witnesses to the text, one has more basic questions to settle first: what was the nature (and date) of the original text? And is our evidence sufficient to enable an intelligent reconstruction of it, or must we be satisfied with, say, the medieval vulgate? An influential argument in recent decades has been that the tradition was completely oral in its early stages, and was not totally stabilised until the second century BC; the extra verses and variant readings found in early papyri and quotations reflect, on this view, an ongoing tradition of composition-in-performance. In such an environment, individual performances, even by the same poet, could vary greatly from each other in their details, while being recognisably instantiations of the same traditional story. The practice of recording such performances in writing grew only slowly, as did the notion of a single, fixed text of 'Homer' (there was no original genius, only a name for the tradition); consequently, what arrived at the library of Alexandria was a series of different recensions, upon which the great scholar Aristarchus succeeded in imposing lasting

---

[20] There is also M. L. West, *Sing Me, Goddess* (London, 1971), a translation of *Iliad* 1 into the metre of the Kalevala (most familiar from Longfellow's *The Song of Hiawatha*), a typically original move. The translation well reflects the rapidity and syntactic simplicity of the original, but the metre has a jauntiness not all readers would think suitable. See J. B. Hainsworth in *Classical Review*, n.s. 23 (1973), 265. Re-reading it myself after many years, I found it much more appealing and successful than I remember thinking when it first appeared. It is not great poetry, but one must admire the easy navigation between faithfulness to the original and natural idiom in English, the Scylla and Charybdis of translation.

[21] All these West inspected personally where possible. The *Iliadic* papyri are all listed in M. L. West, *Studies in the Text and Transmission of the Iliad* (Munich and Leipzig, 2001), pp. 88–138, with catalogue numbers, dates, lines covered, editions, images where available etc.—a small example of his extraordinary thoroughness and industry. For the *Odyssey* papyri, see his edition of the *Odyssey* (Berlin, 2017), pp. xxvii–xlv.

order in the second century BC, at least in respect of the number of lines. If this reconstruction of the early tradition were true, editing an original *Iliad* and *Odyssey* of the seventh century (as West dated them) would be futile; instead, one would attempt an edition that reflects these multiple recensions, putting all the variants and extra verses on equal footing. One can see why this scenario has proved attractive: it takes account of what, on one view anyway, was the state of epic poetry in 'Homer's' day; and secondly, it responds to a suspicion that scholars' inability to agree, after two hundred years of debate about the status of readings in manuscripts and papyri, and the contribution of Aristarchus and his ancient peers, is the sign of a problem misconceived. There are, however, decisive arguments against this theory, some of them put forward by West.[22] The texts of the *Iliad* and the *Odyssey* were certainly recorded in writing at the time of their composition (probably by two different poets, as West consistently argued and most people now believe), and were faithfully preserved thereafter; the variation in early texts can easily be accounted for by the ancient (loose) manner of citation and liberties taken by rhapsodes who recited the poems (as some actors treated the texts of tragedy), and it is clear that all the variants take their bearings from a single point of reference. So, the problem is to find a convincing interpretation of the extraordinarily difficult evidence about the pre- and post-Alexandrian text, and on this basis to decide, line by line and word by word, what one judges the first author to have written. The problem is of a magnitude and complexity greater than any other in classical studies, and has attracted the concentrated attention of many great Hellenists. Nonetheless, even in a field so well ploughed as this, West characteristically found convincing new answers, or new arguments in support of old and neglected ones.[23] The knowledge of the sources and the scholarship is matchless; sovereign judgement is on display every step of the way. The argument is laid out in his *Studies* in crystal-clear prose. It and the editions should be definitive for many years to come.

When it came to constituting the text itself, West fearlessly followed his conclusions where they led him, so that, for instance, he was prepared to designate many more lines of the poems as post-Homeric interpolations than is standard nowadays (in addition to Book 10 of the *Iliad* in its entirety, which most scholars agree is intrusive).

[22] See West, *Studies in the Text and Transmission of the Iliad*, esp. pp. 158–161, various chapters in *Hellenica* 1, and M. L. West, 'Response to *BMCR* 2002.11.15' (i.e. to two reviews of *Studies*), in *Bryn Mawr Classical Review*, 2004.04.17. All the arguments are laid out by B. Currie, *Homer's Allusive Art* (Oxford, 2016), pp. 13–22.

[23] An important foundation was laid by Stephanie West in her book *The Ptolemaic Papyri of Homer* (Cologne, 1967); she went on to write 'the best commentary on any four books of the *Odyssey*' (M. L. West, *The Making of the* Odyssey (Oxford, 2015) p. viii). West dedicated this book to her, and comments that she 'has supported my Homeric and other studies for over half a century'. One can imagine many conversations over the dinner table.

West also took the trouble to update and expand the list of 'testimonia' compiled by earlier editors; these are quotations of Homer in other ancient authors, numbering in their thousands. They are listed in their own register between text and critical apparatus: another chalcenteric labour. This edition of Homer is one of the monuments of Classical scholarship; it is a masterpiece of patient brilliance and one of West's great achievements.

West sets out his theory about the identity of Homer in an important article, 'The invention of Homer'.[24] Briefly, the putative author of the poems was not a real person but the mythical eponym of the 'Homeridai', the guild of professional bards identifiable from the late sixth century onwards who thought of themselves as notionally 'descended' from someone named Homer (like doctors, the Asclepiadai, from Asclepius). West revives an older argument of Marcello Durante that the root *hom-* refers to an assembly of the people, a festival at which epics were performed (like an eisteddfod), so that a projected figure—'Homer'—would be its embodiment. All the traditional poems of which these singers were the guardians were the work of 'Homer', but in the early archaic period most poems were simply anonymous. Though drawing for the most part on views already on offer, West here offers the most cogent explanation of all the evidence, and ably accounts for all aspects of the preservation and subsequent reputation of the poems.

When it comes to the question of how these two poets composed their works, however, West's views are less apt to command assent. It is necessary to explain something of the history of Homeric criticism. Beginning with Friedrich August Wolf in the late eighteenth century, the 'Analytical' school argued that the poems were products of successive compilation of originally independent lays, which could be separated from one another by analysis of their language and content in order to identify earlier and later strata in the poems. Younger dialect forms, for instance, created a presumption of lateness of the passage in question; supposed contradictions in the content of two passages meant they could not both stem from the same hand. The bewildering variety of competing analyses on offer tended to discredit the method, and a 'Unitarian' school, mainly American but also German, arose, which demonstrated beyond question the essential artistic integrity of the poems. Milman Parry's proof in the 1920s that formulaic composition-in-performance was in the background entailed that any given line might combine older and younger dialect forms, since they all belonged to the traditional vocabulary learned (and updated) by each generation of bards. This discovery greatly favoured the Unitarian view, and scholars began to explain the poem's characteristics in terms of the exigencies of live performance.

---

[24] M. L. West, 'The invention of Homer', *Classical Quarterly*, n.s. 49 (1999), 364–82, reprinted in *Hellenica* 1, pp. 408–36. Cf. West, *The East Face of Helicon*, pp. 622–3.

West was a Unitarian, but not an oralist. He fully appreciated the relevance of the oral background to analysing the formulaic language of the poems, and naturally he accepted (sometimes) that many plot motifs and turns of phrase were common bardic stock and thus insufficient to establish specific links between two passages that employ them. On the other hand, he believed that a literate frame of mind took hold relatively quickly in the wake of the *Iliad*, so that the 'rhapsodes' creativity was able to express itself in novel ways: by adding new sections to these texts, transcribing passages from one into another, or making forced combinations of separate pieces'.[25] He spoke habitually of this line or that being 'adapted from' or 'modelled on' or 'copying' another one in our surviving texts, as if the poet might not have heard similar lines in who knows how many performances. Although scholars have recently started to argue that intertextuality can be quite precise even in an oral environment, West's view of the bards' procedures often makes them look much like an Apollonius or Virgil.

Moreover, West thought that the Analysts had drawn attention to many problems in the poems—inconsistencies, inconcinnities of one sort or another—which had never been satisfactorily explained, least of all by the oralists. Not believing in multiple authorship like the Analysts, he posited that the two poets of the *Iliad* and the *Odyssey* had both written their poems down, but had revised and expanded them over the years and decades. The conditions of early writing—the expense of the writing material, the many rolls needed for poems of such length—meant that such revisions would be accommodated not by recopying the whole poem, but by cutting the rolls and pasting in new passages, or by writing shorter changes in the margins. Inevitably, this process resulted in the irregularities and contradictions we now see in the text.

West had already identified some passages in Hesiod's *Theogony* and *Works and Days* as author's insertions.[26] The Homeric case is fully argued in *The Making of the* Iliad (Oxford, 2011) and *The Making of the* Odyssey (Oxford, 2014). In these books, West repeatedly cites the Analysts, fully aware that he would puzzle and infuriate a generation of scholars raised in the oralist faith and taught to despise such antediluvians. The former book opens with a vigorous defence of their (and his) procedures. Now the theory seems perfectly possible; if Homer wrote his poems down, he could have made changes in the manner posited. That authors' revisions have left traces in the manuscript tradition is accepted for some later writers, while another scholar has

---

[25] M. L. West, '*Iliad* and *Aethiopis*', *Classical Quarterly*, n.s. 53 (2003), 13–14, reprinted in *Hellenica* 1, p. 261. On the next page he begins his conclusion 'Once we shake the oralists off our backs.'

[26] See M. L. West, *Hesiod: Works and Days* (Oxford, 1978), p. 58 n. 1; see also M. L. West, 'Is the *Works and Days* an oral poem?', *Hellenica* 1, pp. 146–58. G. P. Goold, 'The nature of Homeric composition', *Illinois Classical Studies*, 2 (1977), 1–34, had the same view about Homer. West diagnoses author's insertions in Euripides at *Hellenica* 2, pp. 302, 311–17, and in his review of D. Mastronarde's *Phoenissae*, *Classical Philology*, 85 (1990), 315–16.

diagnosed alternative versions of the same material in Hesiod's *Works and Days*.[27] The principal difficulty, however, is that the offence felt by Analysts and West too often depends on modern and unexamined common-sense judgements about what is acceptable, expected or 'logical' in matters of archaic composition and style. For all that the texts were written down, the oral, performative environment in which they were composed and delivered must determine both aesthetics and poetics. If in the Embassy scene in *Iliad* 9 there seem now to be two ambassadors, now three, would ancient poets and audiences be much bothered? If it looks as if the Greeks and Trojans will come to blows in Book 2, is it a problem that Agamemnon's crisis of confidence postpones the battle for nine books? If Thetis in her anguish weeps that Achilles will die as soon as he kills Hector (*Il.* 18.95–6), is it a problem that he is not in fact killed right away?[28] Are an unusual present tense and a vague pronoun enough to prove that a passage originally had another home (*Od.* 7.103–31)?[29] Can the supposed 'organic' qualities of an epic poem's episodes (or the lack of them) allow inferences about multi-stage composition? Such questions arise repeatedly, and their answers can have far-reaching implications for one's understanding of the poets' art and for literary history. West knew these poems better than anyone (large tracts by heart) and he made many brilliant observations, which one must take into account. It should not be forgotten that his Unitarian analysis of Hesiod's poems successfully defended the authenticity of many passages on the basis of a better understanding of the poet's procedures. Nonetheless, his strongly anti-oralist perspective is one shared by few scholars, even those who accept the early fixity of these texts.

In rounding off this section on West's studies of epic, a word on *The Hesiodic Catalogue of Women: its Nature, Structure, and Origins* (Oxford, 1985) is appropriate. This poem of the sixth century BC is a revision of a Hesiodic original, a continuation of the *Theogony*, whose subject is divine unions, with an account of the offspring of gods and mortal women. In recounting his genealogies, the poet refers summarily to

[27] L. E. Rossi, 'Esiodo, Le Opere e i Giorni: Un nuovo tentativo di analisi', in F. Montanari and S. Pittaluga (eds.), *Posthomerica*, vol. 1 (Genoa, 1997), pp. 7–22. These alternatives would offer a choice of scripts for recitation on different occasions, which is a slightly different idea but envisages a similar use and subsequent history of the manuscript. In his new Oxford Classical Text of Herodotus, N. G. Wilson has diagnosed author's insertions; for discussion, see his *Herodotea* (Oxford, 2015).

[28] *Hellenica* 1, pp. 251–2. The two events are causally and symbolically linked.

[29] This passage is West's prize exhibit: see M. L. West, 'The gardens of Alcinous and the oral dictated text theory', *Acta Hungarica*, 40 (2000), 479–88, reprinted in *Hellenica* 1, pp. 265–76 ('incontrovertible', p. 268), and *The Making of the* Odyssey, pp. 94 ('irrefutable'; the intrusion of present tenses is 'wholly unparalleled and unconscionable'), 132, 188. Contrast I. de Jong, *A Narratological Commentary on the* Odyssey (Cambridge, 2001), pp. xiii (under 'description'), 176 for the generic norm that is here not observed and the reason why. To explain a new problem arising from his analysis, West had to advance a secondary, ad hoc hypothesis (*The Making of the* Odyssey, p. 188 n. 65).

the great exploits associated with these offspring; he thus offers a poetic history of the heroic age of Greece, down to the Trojan War and its immediate aftermath. Knowledge of the poem gradually improved in the course of the twentieth century with the appearance of papyri, but leapt forward with the publication of volume 28 of *The Oxyrhynchus Papyri*, containing nearly as many papyri again as had hitherto been found; a further substantial fragment appeared in 1981. These discoveries enabled the edition of Hesiod's fragments jointly published by West and Merkelbach in 1967, and provided the basis of a reconstruction of the *Catalogue*, begun by Merkelbach and finished by West. This poem was, among other things, the well-spring of the mythographical tradition in antiquity. As the commentary on the *Theogony* had included a worldwide survey of theogony, and the commentary on the *Works and Days* a similar survey of wisdom literature, the *Catalogue* book collects examples from all over the globe of genealogical poetry. Anyone who has tried to find their way through the bewildering morass of variants that is Greek mythology has reason to be grateful for this superb book. It will still take a determined effort to work through West's treatment—that lies in the nature of the subject. But diligence will be rewarded not only with enhanced admiration for this early poet's feat, but with an appreciation of the cultural and political importance of genealogies in archaic Greece. As West patiently maps both the wood and trees of this unusually dense forest, one cannot but marvel at his skill, learning and powers of combination. This is another work of genius.

*****

To move on to lyric poetry, early in his career West produced *Iambi et Elegi Graeci ante Alexandrum cantati* (*IEG*), in two volumes (Oxford, 1971–2); the advent of the 'New Simonides' and other finds necessitated a second edition (1989–92).[30] His qualities as an editor are especially conspicuous in these volumes: superb philology; clarity of layout and ease of consultation; judicious selection of supporting materials; brilliance in conjectural emendation of manuscripts and supplementing lacunae in papyri. The last quality irritated more conservative critics. Bruno Gentili, co-editor of the Teubner edition of the elegiac poets, objected to West's practices, challenging also his use of evidence in assigning fragments and reconstructing poems.[31] He was particularly

---

[30] The 'New Simonides', sc. Oxyrhynchus Papyrus 59.3965 published in 1992, was one of the more spectacular papyrus discoveries of the late twentieth century (its overlap with *POxy* 22.2327 enabled that papyrus, hitherto anonymous, also to be identified as Simonidean). West made a decisive contribution to the understanding of the text and its place in Simonides' oeuvre, both in the first publication and in his 'Simonides Redivivus', *Hellenica* 2, pp. 111–18 (originally published 1993). For studies, see D. Boedeker and D. Sider (eds.), *The New Simonides: Contexts of Praise and Desire* (Oxford, 2001).

[31] B. Gentili, *Gnomon*, 52 (1980), 97–101.

exercised by what West had done with the gibberish transmitted in the single manuscript that preserves Semonides fr. 10a:

καὶ μήτ᾽ ἀλλ᾽ οὕτως γὰρ ἂν εὖ μεθ᾽ ὕδωρ θαύμαζε μὴ δὲ [. .]ύρη γενειάδα· μηδὲ ῥυποχίτων ἔση ἕν τε χώρα

which West turned into:

καὶ μήτ᾽ ἄλουτος γαυρία σύ, μήτ᾽ ὕδωρ
θαύμαζε, μηδὲ [κο]υρία γενειάδα,
μηδὲ ῥύπωι χιτῶνος ἔντυε χρόα.

and translated as:

Do not be proud of never washing, nor
a water-maniac; grow no bushy beard,
nor wrap your body in a filthy cloak.

In the apparatus West notes simply 'correxi', which as he well knew should be used only when an editor thinks no doubt can be entertained about what he has printed. Gentili gave this as his principal example of the work's 'most obvious limitation, West's very conception of a critical edition as a bravura arena in which to parade one's skill in inventive conjecture'.[32] Another reviewer, however, more of West's way of thinking, closed by remarking 'one must sincerely raise one's square [hat] to salute a scholar who can dig this out of a corrupt παράδοσις'.[33] Douglas Gerber and Bruno Snell both accepted the reading in their editions. West was as capable as anyone of defending an unjustly impugned paradosis, but like Housman he had no tolerance for critics who placed unreasoning faith in the manuscripts. He expounded his procedures with his usual force and clarity in the near contemporary *Textual Criticism and Editorial Technique* (Stuttgart, 1973), a book that teems with excellent practical advice on every aspect of the craft, and which has done good service for many, not just for classical editors. Whatever one may make of his bolder readings, they are unfailingly intelligent, often arresting and always thought-provoking. We shall say more of this when considering the edition of Aeschylus.

Critics complained with greater justice that the information offered in the apparatus of *IEG* was too sparse, that testimonia to the authors' lives and works were all but omitted, and that West was less scrupulous than he might have been in reporting other scholars' conjectures. In keeping with the editorial policy of the Teubner series, Gentili and his fellow editor Carlo Prato gave the reader far more in the way of testimonia,

[32] Ibid., 97: 'Il suo limite più evidente è nell' idea stessa che il West ha dell' edizione critica, concepita come una palestra di bravura, in cui dar prova di capacità inventiva nel congetturare.'
[33] G. Bond, *Classical Review*, n.s. 25 (1975), 181. West first proposed the reading in *Maia*, 20 (1968), 196. The paradosis is the tradition as received in the manuscripts.

parallels, conjectures and bibliography, so that the edition was also a basic commentary. One is often grateful for such information, even if it makes consultation less easy. It can be overdone. In reviewing Gentili–Prato, West was relatively mild in his criticism,[34] but in the witty choliambic verses he wrote as prefaces to his own volumes (who else but West would do that?) he permitted himself a fairly sarcastic barb, partly for the sake of an ingenious pun on their names.[35] Specialists need to consult the fuller texts, and Gerber's Loeb edition with its translations is the first port of call for students and non-specialists, but the qualities of *IEG* make it for many scholars the working edition to have to hand.

As a companion volume to *IEG* West published *Studies in Greek Elegy and Iambus* (Berlin and New York, 1974), as he was later to publish similar volumes of studies to accompany his editions of Aeschylus and the *Iliad*. Here one finds his groundbreaking study of the genres of elegy and iambus (the latter including the controversial argument that Lycambes and other figures in iambic poetry were not real people, but traditional types); his account of the history of the corpus attributed to Theognis (contested in details, but the general idea of a multi-stage development is widely accepted); his argument that Theognis lived in the latter part of the seventh century BC instead of the sixth (which has convinced almost no one); a brief account of the life and works of Mimnermus; and a commentary on selected passages. One will also find a dense and extremely rewarding chapter on the language and metre of the poets, a treatise that any aspiring philologist should study with the utmost care.

The principal texts in *IEG* were also published as an Oxford Classical Text;[36] the genuine Theognis (in West's judgement) from the Theognidea, along with the fragments of Demodocus, the hexameters of Phocylides (who, having written no elegies, was omitted from *IEG*) and anonymous hexameter gnomic fragments were published as a Teubner text.[37] The edition of the *Anacreontea* has already been mentioned; it was the first truly critical edition since the early twentieth century, and at once became standard.[38] The preface provides a concise discussion of sources, language and metre, history of the collection and scholarship on it; the text itself is equipped with a useful

---

[34] *Classical Review*, n.s. 31 (1981), 1–2.

[35] 'Strictius, puto, nullo / gramina videbis rasa gentili prato, / nec ubi redundat usquequaque laetamen': 'you shall not see, I think, the grass more closely mown on any foreign meadow, nor the fertiliser flooding everywhere' (*laetamen*: translate 'manure' if you choose). M. L. West, *Iambi et Elegi Graeci ante Alexandrum cantat*, 2nd edn, vol. 2, (Oxford 1972), p. 1.

[36] M. L. West (ed.), *Delectus ex Iambis et Elegis Graecis* (Oxford, 1980).

[37] M. L. West (ed.), *Theognidis et Phocylidis Fragmenta et Adespota quaedam gnomica* (Berlin and New York, 1978). The latter was in the Teubner 'Kleine Texte' series and so did not include the wealth of additional material mentioned, but it did have illustrative parallels and brief explanatory comments.

[38] David Campbell's 1988 Loeb edition is excellent and probably more widely used, given the translation and the price; but naturally it builds on West, adopting twelve of his emendations.

collection of parallel passages. Although he never edited the Lesbian poets nor those in Page's *Poetae Melici Graeci*, West made dozens of contibutions in reviews and articles to the constitution and understanding of their texts. Editions of papyri routinely reported his suggestions, since any editor with sense approached him for advice pre-publication. Even so, West would often contribute his own separate treatment in an article in the *Zeitschrift für Papyrologie und Epigraphik*, usually in the next number after the edition appeared. Mention should also be made of *Greek Lyric Poetry* (Oxford, 1993), in which he translated all the poems and fragments of iambic, elegiac and melic poets, except for Pindar and Bacchylides, down to 450 BC. This is the most successful of his translations, and with its concise notes and introduction is a superb entry point to the period for students and general readers. One of his last contributions in this field, 'Pindar as a man of letters' (*Hellenica* 2, pp. 129–50), documents the breadth of Pindar's knowledge of earlier literature, and cogently argues that he must have derived a great deal of it from reading books. However one draws the balance between oral and literate in the days of Homer, scholars have consistently ignored the high degree of literacy among the intelligentsia by the sixth century BC if not before. West's article is a firm refutation of this error, traceable to Eric Havelock's influential *Preface to Plato* (Oxford, 1962), and imparts momentum to a re-evaluation of this topic now under way. The oral, performative environment had been neglected before Havelock, and we have learned much from this line of research; but the recognition that there was simultaneously a wider literate environment (literary, indeed, in a strong sense) makes a great difference in how one assesses matters such as intertextuality, real and implied audiences, the author's persona and so on.

The edition of Aeschylus' surviving plays naturally holds pride of place in West's oeuvre on tragedy. The paradosis presents many problems of exceptional difficulty, resulting partly from the tragedian's bold use of language, partly from the state of the manuscripts. *Prometheus Bound* offers its own set of peculiarities, sufficient to have convinced West and many others that it is not by Aeschylus (that it was written by his son Euphorion is an older suggestion developed by West with further arguments).[39] The edition is based for the first time on a complete knowledge of the manuscript tradition (all MSS down to and including the fourteenth century were consulted). The relationships between the manuscripts is much better understood, and the manner of reporting them (and editorial interventions) in the apparatus is at once more informa-

[39] The full title of the edition is *Aeschylus: Tragoediae cum incerti poetae Prometheo* (Stuttgart, 1990); West is the first editor to condemn the *Prometheus Bound* explicitly in his text. See M. L. West, 'The Prometheus trilogy', *Hellenica* 2, pp. 250–86; M. L. West, 'The authorship of the Prometheus trilogy', in *Studies in Aeschylus* (Stuttgart, 1990), pp. 51–72; and M. L. West, '*Iliad* and *Aethiopis* on the stage: Aeschylus and son', *Hellenica* 2, pp. 227–49 (in which he suggests further plays transmitted under Aeschylus' name that may have been written by his son).

tive and economical. The register of testimonia has been improved, as in the edition of Homer. The Praefatio, in West's fine Latin, discusses the paradosis, matters of grammar, prosody, orthography; a brief section clarifies the contributions of the six-teenth-century scholars Auratus (Jean Dorat) and Franciscus Portus. An appendix provides the editor's analysis of the metres of all lyrical portions. The accompanying volume of *Studies* discusses the manuscripts and the history of Aeschylean textual criticism at greater length, including the editor's calculation of who had made the greatest number of successful emendations; far out in front is Turnebus in the six-teenth century (191), followed by Gottfried Hermann in the nineteenth (135). West himself contributed many conjectures; he also performed signal service in resurrecting neglected ones. In 'Forward into the past', he singles out the obscure K. H. Keck's brilliant restoration of *Suppliants* 599 (σπεῦσαι. τί τῶνδ' οὐ Διὸς φέρει φρήν, for the manuscript's σπεῦσαί τι τῶν δούλιος φέρει φρήν), ignored by everybody save J. Oberdick in his edition of 1869, as emblematic of the textual critic's art.[40] Another striking example of resurrection is Badham's overlooked 'Hekate' (Ἑκάτα) for the manuscripts' odd 'the fair one' ((ἁ) καλὰ) in *Agamemnon* 140. The *Studies* also con-tains chapters on the Prometheus and Lycurgus trilogies, and on 'The formal struc-tures of Aeschylean tragedy'. The underlying schema imputed to Aeschylus in the latter is too general, and rests upon too small a sample (the four early plays, in which already there are significant variations) to command assent, but the observations on individual structural elements are often instructive.

The heart of the *Studies* is some 240 pages of notes on selected passages. Aeschylus is a kind of Everest of Greek textual criticism, and in the nature of things conjectural emendations are less likely to find acceptance than in other authors. By my count West made 339 conjectures (excluding the orthographica flagged up in the Praefatio), of which he admitted 119 into the text. These can be dazzlingly right—they are often original, ingenious and elegant; boldness is a common feature. But there are many times when one wishes he had considered matters a little more carefully and resisted the temptation, born of his amazing facility, to emend so much. Were all 119 correc-tions to be correct, West would be, on his own reckoning, the third most successful emender of Aeschylus in history. Or even the most successful: Turnebus in the Renaissance could pluck the low-hanging fruit, and so much work has been done since Hermann that the chances of this number of new conjectures being right would seem small. Another editor would mark more passages as incurable, and confine

---

[40] *Hesperos*, p. xxii, where he translates the paradosis as 'he is able to execute deed as soon as word of whatever his servile mind brings forth'; the previously favoured emendation (βούλιος for δούλιος) translates as 'of whatever his counselling mind brings forth'; Keck's emendation means 'he is able to execute deed as soon as word. What of these things is not brought forth by Zeus' mind?'

conjectures to the apparatus; the older Oxford edition of Denys Page (Oxford, 1972) is safer in this sense (and cheaper), and will probably be the most commonly used. West knew of course that a reading's rightness was normally a matter of probability, not certainty, but he placed the bar for admission much lower than other scholars would; the criticism levelled against *IEG* is more justified in this case. The idea, however, that minimal intervention was the more responsible course was forcefully and eloquently rebutted on several occasions by West; in respect of Aeschylus, see *Studies*, pp. 369–72. In this as in much else, Fraenkel was the link between his teacher Wilamowitz and West.

Apart from the Aeschylean work, West also produced an edition of Euripides' *Orestes* in the Aris & Phillips series (equipped with a translation and aimed primarily at students, but there are many notes valuable also to scholars and the usual clutch of emendations). One should mention also the series 'Tragica', seven articles of notes published in the *Bulletin of the Institute of Classical Studies* between 1977 and 1984 (excerpted in *Hellenica* 2).

*****

Hellenica 3 opens with eleven chapters on Greek philosophy, or more precisely Greek cosmology, the dominant topic among early Greek thinkers. In the other branches of philosophy that preoccupied later Greek philosophers West had no interest, at least not such as provoked publication. His abiding interest in astronomy is on display here (note especially 'The midnight planet', pp. 110–15). The chapter 'Alcman and Pythagoras' stands out as a rich exposition of cosmological thought in the most unlikely of places, a seventh-century BC Spartan poet. But the most insistent theme in these essays—appearing also in several chapters in the rest of the volume—is the Near Eastern background of Greek thought. In addition to his many articles, there are three books: *Early Greek Philosophy and the Orient*; *The Orphic Poems*: not primarily focused on the interface of the two cultures, but frequently having occasion to discuss it); and the mighty *The East Face of Helicon*.

The first of these, which is dedicated to Walter Burkert, makes out the thesis that there was a period of intense Iranian influence on Greek thought between about 550 and 480 BC (with some influence before, but none after until dialogue resumed in the fourth century). He attributed the influence primarily to meetings between Greek thinkers and wandering Magi. There are extended discussions of Pherecydes of Syros, Anaximander and Heraclitus, in which the connections are explored in detail. The book was roughly handled by critics, both for its general thesis and in its treatment of the texts. In response many years later, West commented that the book leaves a good deal to be desired, as others have noted. But it has aroused enthusiasm in some quarters, and I do not disown it entirely. The accounts of Pherecydes and Heraclitus are perhaps the most substantial contributions in it. I regret that the book has been almost totally ignored, *totgeschwiegen*, by the 'professional' historians of Greek philosophy, who remain absorbed in their own agenda.[41]

He clung to the idea of Iranian influence in a late article on Zoroastrian influence in Greece,[42] which suggests that he accepted the criticism in respect of the details, but not of the general thesis. However, the more extended, if implicit, response was *The East Face of Helicon*. Although this book concentrates on myth and literature to the deliberate exclusion of science and philosophy, by documenting the myriad parallels between the literatures and mythologies of the Near East and those of Greece (in which, after all, many of the cosmological ideas were embedded), and by suggesting many ways such motifs and ideas might have crossed cultural boundaries, from the Bronze Age onwards almost without break, West implicitly concedes that the narrow window of influence in philosophy 550–480 BC is a reflection only of the surviving texts and part of a wider picture. He acknowledges that the routes of transmission were complex and mostly beyond our detection (though he still stresses the role of wandering wise men—poets rather than Magi in the case of literature).[43] One may detect too a mellowing of the pioneer's enthusiasm, in that the earlier book effectively claimed that the Near East deserved the credit for the brilliant innovations of early Greek philosophy:

> But what invaded Greek speculation in the mid sixth century was no mere convolvulus that withered away when its season was past, leaving the sturdy stems of Hellenic rationalism to grow unimpeded as they had always meant to. It was an ambrosia plant, that produced a permanent enlargement where it touched. In some ways one might say that it was the very extravagance of oriental fancy that freed the Greeks from the limitations of what they could see with their own eyes: led them to think of ten-thousand-year cycles instead of human generations, of an infinity beyond the visible sky and below the foundations of the earth, of a life not bounded by womb and tomb but renewed in different bodies aeon after aeon. It was now that they learned to think that good men and bad have different destinations after death; that the fortunate soul ascends to the luminaries of heaven; that God is intelligence; that the cosmos is one living creature; that the material world can be analysed in terms of a few basic

---

[41] West, 'Forward into the past', p. xxvii. Reviews: *Classical Review*, n.s. 24 (1974), 82–6 (G. S. Kirk); *Gnomon*, 47 (1975), 321–8 (M. Marcovich); *Mnemosyne*, ser. 4, 32 (1979), 389–96 (W. J. Verdenius).

[42] West, 'The classical world'.

[43] '[W]e can no more count and describe the sources of all the eastern motifs and procedures than plot the flow of waters beneath the surface of a marsh'; West, *The East Face of Helicon*, p. 629.

constituents such as fire, water, earth, metal; that there is a world of Being beyond perception, beyond time. These were conceptions of enduring importance for ancient philosophy. This was the gift of the Magi.[44]

The later book speaks of steady influence rather than a quasi-magical transformation, and contemplates scenarios of extended, complex interaction. The nuancing also blunts the Orientalism of the passage just quoted. These refinements apart, however, West re-emphasises in the strongest terms that the Near East contributed decisively and massively to Greek culture. The astonishing number of parallels, he argues, could yield no other conclusion, even if any individual parallel might not entail a causal relationship. The topics he raises in the peroration of the earlier book, and those in *The East Face of Helicon*, still demand assessment, and a new generation of scholars is rising to the challenge. Even where they disagree with some of his findings (the force of the cumulative argument in particular is coming under pressure), this shift in orientation acknowledges the impact of West's work. Classicists are starting to learn the languages: West himself learned Hittite, Akkadian, Ugaritic, Phoenician, Aramaic and Hebrew in order to write this book.

The Orphic Poems is another groundbreaking work. It is the best synoptic treatment of the whole Orphic tradition. There is no underlying unity in this tradition, so what West attempted instead was to establish when and by whom Orphic texts were written, and how they related to one another. The stemma he drew (see p. 264) was very complex, and one may think that, like many manuscript traditions, this one is too contaminated by cross-fertilisation between branches to yield a stemma. In 'Forward into the past' West defends the book against charges of excessive speculation and reiterates his belief in the soundness of the reconstruction. But whatever one may make of the stemma, there is no doubt about the importance of this work for its discussion of many problems.

The Derveni papyrus was a cardinal witness, and since West's role in the first publication has been a matter of speculation, it is worth reporting on it here. This remarkable text, an extremely idiosyncratic commentary on an early Orphic theogony, was first excavated in 1962. Seven of its surviving twenty-eight columns were published in 1965. West copied what he could from the papyrus itself in the museum in Thessaloniki in 1970, which added four more columns and some smaller fragments to what was known. The Greek editors made a full draft text available in 1980, which circulated in photocopies. West, who had acquired one from Eric Turner FBA (1911–1983), shared it with Walter Burkert. Regarding the situation as intolerable, Burkert persuaded Merkelbach, the editor of the *Zeitschrift für Papyrologie und Epigraphik*, to publish it without proper permission. It appeared anonymously in 1982, at the back

---

[44] M. L. West, *Early Greek Philosophy and the Orient*, (Oxford, 1971) pp. 241–2.

of volume 47, separately paginated and unrecorded in the issue's table of contents.[45] The breach of scholarly protocol caused ill feeling, but most readers thought it was justified, given the importance of the text and the unconscionable delays. The full edition did not appear until 2006, although before then full materials were made readily available by the Greek editors to those who asked.

Music and metre are the principal topics occupying the remainder of *Hellenica* 3. On the first West published *Ancient Greek Music* (Oxford, 1992) and, with Egert Pöhlmann, *Documents of Ancient Greek Music* (Oxford, 2001), an edition and commentary on the surviving texts. These are, once again, works of fundamental importance. The monograph transformed the subject, not only patiently explaining the daunting technicalities but also providing detailed research on instruments, singing, the role of music in Greek life and the history of musical developments in antiquity. *Greek Metre* (Oxford, 1982) is the standard reference.[46] West made many refinements to the empirical system of analysis worked out by Wilamowitz and Paul Maas (1880–1964), which reflects ancient reality far better than any other, despite some claims to the contrary. The book is ordered chronologically, to make the point that metre is an aspect of literary history and, where possible, to explore aesthetic implications; a typically ingenious glossary-index helps those seeking a synchronic perspective and offers the best brief guide available to the formidable jargon of this topic.

A subject spanning all three volumes of *Hellenica* and all stages of West's career is the Indo-European heritage of Greek literature. This interest culminated in the masterpiece *Indo-European Poetry and Myth* (Oxford, 2007). Here West deployed his knowledge of yet more languages (some of which he had known since his youth). 'Most translations offered for quoted passages (Vedic, Avestan, Greek, Old Norse, etc.), are my own', he writes (p. xiii). Who knows how many languages are covered by that 'etc.', but it certainly included Germanic languages (Old English, of course),[47] and a smattering of Celtic and Slavic. The twelve chapters explore a long list of common themes, which comprise many secondary topics, all richly illustrated with examples; the bounty of this treasure house is practically beyond measure. One may query (West himself queries, p. 24) how to gauge the force of mythological parallels (as opposed to more precise linguistic connections), how many and what kind of parallels between different branches it takes to certify something as Indo-European, and in what ways exactly the inheritance we can document from our vantage point

---

[45] For details, see W. Burkert, 'The true story of the anonymous edition', in I. Papadopoulou and L. Muellner (eds.), *Poetry as Initiation* (Washington, DC, 2014), pp. 113–14; cf. M. L. West, *The Orphic Poems* (Oxford, 1983), p. v.

[46] There is also the shorter M. L. West, *Introduction to Greek Metre* (Oxford, 1987).

[47] Verses of the ninth-century monk Otfrid of Weissenburg are translated from his particular dialect of Old High German at ibid., p. 190.

had real traction in any given case (a wider problem, pertinent to any study of tradi-tions). These are questions that will continue to be debated; as in *The East Face of Helicon*, West has given us the tools to do it. The book closes with an 'Elegy on an Indo-European hero', an affecting ode incorporating many of the motifs identified as common inheritance.

Finally, two books that were parerga of the Indo-European labours: *The Hymns of Zoroaster: a New Translation of the Most Sacred Texts of Iran* (London, 2010), and, even more impressively, *Old Avestan Syntax and Stylistics: with an Edition of the Texts* (Berlin and Boston, 2011).[48] The syntax book usefully filled a gap in the literature, and specialists need to consult it. On the other hand, these texts are extremely problematic, and no agreement exists on many questions of reading, construction and meaning. West's critical edition offers thirty-three emendations, of which nineteen are put in the text; the translation naturally depends on these, and the syntax book had also to take a view on the construction of controversial passages. The introduction to the transla-tion fully acknowledges the existence of many serious difficulties, but in the course of the book the author signals only 'the most major uncertainties' (p. viii). This policy has drawn criticism, and the tendency to disregard, or even denounce, trends that have dominated criticism in recent decades did not go down well in some quarters.[49] Yet, if one takes the view that scholarly progress depends on crossing disciplinary boundaries, one can only be grateful for someone who can conduct both sides of a conversation with such fluency. Few people when treading on alien ground would score so well in both accuracy of detail and command of the general picture.[50] Moreover, the book certainly succeeded in its stated aim of making these great texts available to a wider audience.

*****

We have reached the end of this necessarily long overview of the major publications. West's seemingly limitless philological prowess is obvious; it was an ability such as the world rarely sees. No one knew more Greek than he did, and no Hellenist has mas-tered so many other languages. The wonderful facility in Greek and Latin verse

---

[48] 'At that point', commented Robert Parker in his All Souls memorial address, 'even seasoned West-watchers had to gasp'.

[49] West discusses his principles, and explains many of his conjectures, in 'On editing the *Gathās*', *Iran*, 46 (2008), 121–34. The translation was reviewed by D. Durkin-Meisterernst in the *Journal of the Royal Asiatic Society*, ser. 3, 21 (2011), 379–81, and the syntax book by R. Schmitt, *Kratylos*, 57 (2012), 161–70. For West's combative attitude to literary theory, see below p. 116.

[50] Note the review by the Assyriologist A. R. George of *The East Face of Helicon* in *Classical Review*, n.s. 50 (2000), 103–6.

composition was part of this linguistic gift.[51] Perhaps unexpectedly, he read slowly, but he remembered everything; the capacity of his memory seems scarcely credible. It was coupled with an ability to marshal in orderly array the millions of facts stored within, and to present them to the reader with superb lucidity, often deploying ingenious methods of arrangement in which he took special delight. If scholars worship on the altar of Akribeia, the goddess was surely well pleased with this acolyte's meticulous devotions. It is rare even to find a misprint in West's writings. Also obvious is the amazing quantity of publications; extraordinary discipline, concentration and powers of recall must be supposed to account for such an output. The editor of the Aris & Phillips series, Christopher Collard, said that the *Orestes* commentary was the fastest ever produced. There were also scores of encyclopaedia articles and a multitude of reviews. Such contributions are one example of West's generosity, which many young scholars and visitors to Oxford experienced (including the present writer on many occasions). He was responsible, indeed, as an assiduous Dean of Visiting Fellows at All Souls, for bringing many overseas classicists to the College and further developing that successful and important programme.

West was a master stylist in prose, with a gift for happy metaphor, choice diction and graceful rhythm. His virtues as a translator have already been mentioned. Humour is apt to irrupt at any time, beginning with eye-catching titles such as 'Grated Cheese Fit for Heroes', 'Greek Poetry 2000–700 BC', 'Seventeen Distorted Mirrors in Plato', 'Two Lunatic Notes', 'Conjectures on 46 Greek Poets', 'The Way of a Maid with a Moke', 'Ringing Welkins', 'A Vagina in Search of an Author'; a seminar in Oxford about a newly discovered poem of Archilochos of Paros, in which the poet recounts a steamy sexual encounter, was billed as 'Last Tango on Paros'. Among many passages one could quote, I offer this from the tribute to W. W. Cruickshank:

> Claiming to have solved the riddle of the universe is commonly a symptom of schizophrenia. What far-reaching suspicions must he then court, who bids to solve the riddle of *a hundred and eighty-three universes*! Yet such is the extravagant duty appointed for the inquirer whose ambition of the moment is to make sense of the strange cosmology of Petron of Himera. Shall we venture upon the perilous task? Shall we expose to the sniper our reputation, such as it is, for equilibrium? Of course we shall. All must be risked for science.[52]

Non omnia possumus omnes. At the beginning of 'Forward into the past', West distinguishes 'three different approaches to the study of literature':

---

[51] Apart from examples noted elsewhere in this memoir, see *Hellenica* 1, pp. 148–9; 2, pp. 391–3; 3, p. 134; other poems cited at *Hellenica* 3, p. 498. Of modern languages, he spoke German fluently and Italian passably.

[52] *Hellenica* 3, pp. 134–5.

The three approaches are, firstly, consideration of the intrinsic qualities of literary works, their beauties or infelicities, the author's imaginative universe, his compositional habits and techniques, and so on; secondly, inquiry into the work's relationship to the world outside itself, its dating, its authenticity, its debts to earlier models or more loosely to the tradition in which it stands, the intellectual and cultural influences operating on the author; and thirdly—an approach which may draw on both the other two, among others—the endeavour to resolve doubts at the verbal level about what exactly the author wrote and what exactly he meant. These three approaches may be summed up as literary criticism, literary history, and philology.

He identifies himself as a literary historian and philologist. One can of course find remarks throughout his oeuvre about beauties, infelicities, compositional techniques and so forth, and views on such matters are often relevant to textual criticism. He had an unsurpassed sense of Greek literary style; that, his own beautiful English and his various creative endeavours more than justify his calling himself 'an artistic spirit' in the preface to *Hellenica* 3. But his claim here not to be a literary critic did not spring, or spring only, from a sense of limitation or lack of interest; he was hostile to much of the literary criticism practised in the academy, as numerous remarks reveal (in private, he called it 'gush'). So, for instance:

> Of this mass of manuscripts a large part has still not been collated, five hundred years after the invention of printing; if scholars had devoted as much effort to this basic research as they have put into writing 'interpretations' of tragedy, we should be further ahead.[53]

Or:

> structuralism, one of the bulkier bandwagons at present cluttering the road to truth.[54]

Or:

> I disagree fundamentally with those modern scholars who claim that the prophet's style is deliberately esoteric and encrypted, full of intentional double or multiple meanings. In my view, where different interpretations of a sentence are possible, it is the job of the translator or commentator to try to determine which one corresponds

---

[53] M. L. West (ed.), *Euripides: Orestes* (Warminster, 1987), p. 42.
[54] *Hellenica* 1, p. 22, from a review of N. Austin, *Archery at the Dark of the Moon: Poetic Problems in Homer's* Odyssey (Berkeley, CA, 1975). In his Balzan acceptance speech (above, n. 9) West wrote: 'When I received the news of the award two months ago, it came as a total surprise, especially to one who cannot claim to have developed the study of Classical Antiquity in any previously unknown direction, or enriched it with any new concepts or methodology. I practise a style of philology that I learned forty years ago and have seen no reason to change; set in my ways from an early age, I have ignored the changing fashions of scholarship and slept through the noise of the bandwagons that pass in the night. I have from time to time asked new questions and explored neglected fields, but whenever I have done so, I have used traditional procedures.' He describes his work as being of 'a basically old-fashioned kind'.

to the author's intention. To credit him with deliberate ambiguity or multivalence is merely an excuse for indecisiveness, or for showing off the commentator's resourcefulness.

> In choosing between possible interpretations the best guide is contextual coherence. The translator must try to identify the essential thought underlying each sentence— what it is that Zoroaster is wanting to say and striving to express in metrical form—and to trace the sequence of his thinking from stanza to stanza. The more coherent the sequence of thought that can be elicited, while interpreting the words in as unforced a way as possible, the more likely it will be that we have reached a correct understanding.[55]

Or:

> The Greek poetesses, and Sappho above all, set us a challenge, a challenge to be aware of ourselves. Let us by all means play with these figures in our romantic fantasy and seek in their lives things that we miss in our own. Let us bring them into our feminist essays or our erotic fiction as we like. Or let us investigate them as serious scholars and make the effort to interpret their verses and their lives correctly. Only let us be aware of ourselves and recognize which of these things it is that we are wanting to do; and whoever is striving for true knowledge, let him hold on to the principle of always trying to see things as they are, and not as we would wish them to be.[56]

I pass over 'let him' without comment, and the insulting equivalence of feminist essays and erotic fiction. A literary critic would regard a distinction between 'serious scholars' seeing things 'as they are' and others offering feminist (or whatever) readings in the same way a philologist would gape at an unmetrical conjecture or elementary mis-translation. The number of questions begged in these comments about how meaning is produced and received (especially from alien cultures) is large. Another passage in point, too long to quote, is the broadside against 'that curse of contemporary Aeschylean criticism, the belief in the structural significance of recurrent imagery' in West's review of A. F. Garvie's commentary on Aeschylus' *Choephori* (*Hellenica* 3, pp. 223–6, once again a deliberate reprint to underscore the point). One may disagree with Garvie's treatment, but the role of imagery in ancient poetry (and its abundance in Aeschylus) needs systematic, not common-sense, thought; and anything one says is an 'interpretation'. What counts as coherence, and how much it matters, as I have suggested above in connection with Homer, varies from person to person and much more from culture to culture. To overlook such considerations is precisely not to be

---

[55] M. L. West, *The Hymns of Zoroaster: a New Translation of the Most Sacred Texts of Iran* (London, 2010), p. 35. 'In my view ... resourcefulness' was defiantly reprinted as an *obiter dictum* in *Hellenica* 3, p. 488.

[56] *Hellenica* 3, p. 335 (originally 1996). From the same essay, p. 315 n. 1: 'There are several anthologies and general treatments of the more significant poetesses, mainly of a feminist-dilettante nature.'

'aware of ourselves'. One sometimes detects in West's work a kind of hyper-rationalism, fuelled by unquestioning self-confidence, which regards all problems as solvable given sufficient evidence and ingenuity; it is impatient of the messiness of phenomena and expects to find everywhere the happy property of textual criticism that there is but one right answer (the 'truth'). Would it were so.

There were some literary critics West admired; R. P. Winnington-Ingram was one, for whom he wrote the biographical memoir.[57] Winnington-Ingram had, in West's view, the right combination of sensitivity to the text and caution about fashionable approaches; on p. 594 of the memoir West quotes him on structuralism and imposing one's own views on a text, remarks that are echoed in some of the excerpts I have quoted herein. One point on which many readers (certainly this one) would agree with West is the pointless obscurantism of much modern criticism, which he attacked mercilessly in reviews. *Non fumum ex fulgore, sed ex fumo dare lucem*. For the most part, West stuck wisely to his last; his forays into purely literary-critical territory, such as the analysis of Aeschylean structures mentioned above, or the general assessment of the *Orestes* ('a rattling good play'), were few.

West's attitude to literary criticism was widely shared among classicists of his generation. Readers of this memoir will assess it as they will, as they will have different views about the merits of his bold speculations and textual emendations. When all is said, there can be no doubt that Martin West was a very great scholar, indeed a genius, comparable with the greatest of any age. There was no more famous philologist; his name was and is ubiquitous in the professional literature. The citation for the Kenyon Medal justly called him 'the most brilliant and productive Greek scholar of his generation, not just in the United Kingdom, but worldwide' and 'in a class entirely of his own'. His memorial event was attended by hundreds of people from all over the globe. His books were translated into Italian, Greek, Hungarian, Portuguese, Polish. Honours accumulated: degrees from Urbino and Cyprus; memberships of foreign academies and societies (Academy of Athens, Accademia Nazionale dei Lincei, Academia Europaea, Akademie der Wissenschaften zu Göttingen, American Philosophical Society, Hungarian Society for Classical Studies); the Runciman Award for *The East Face of Helicon*;[58] the International Balzan Prize in Classical Antiquity;[59] the British

---

[57] M. L. West, 'Reginald Pepys Winnington-Ingram 1904–1993', *Proceedings of the British Academy*, 84 (1994), pp. 579–97. Winnington-Ingram's important work on ancient Greek music was another reason West approved of him, and probably why he was asked to write the memoir.

[58] The award, named for Sir Steven Runciman CH FBA, is made annually by the Anglo-Hellenic League for a work in English on Greece or Hellenism. West's winning was not without controversy, as some readers thought the book called into question the originality of the 'Greek miracle'.

[59] West used part of the prize money to donate a charming fountain in the Fellows' Garden at All Souls; there was also a substantial donation to Balliol College.

Academy's Kenyon Medal (at the relatively young age of sixty-one); first Emeritus Fellow, then (a rare accolade) Honorary Fellow at All Souls College; honorary fellowships also at Balliol, University and St John's Colleges; and to crown all, the Order of Merit in 2014, joining the *numerus clausus* of twenty-four individuals honoured personally by the sovereign for great distinction in their fields.[60] Invitations to lecture came frequently. Regrettably, he did not keep a list of those accepted, but his report to the Warden for the year 1992 (preserved among the papers at All Souls) mentions talks in Venice, Budapest, Princeton and New York; that for 1997–9 records lectures in Harvard, Cagliari, Tvärminne, Göttingen, Jerusalem, Tel Aviv and Beersheba; for 2001–3, Toronto, Union College (Schenectady), Cornell, Droushia (Cyprus).[61] He was a visiting scholar at Harvard for a semester in 1967–8, at the University of California, Los Angeles, for a quarter in 1986 and did a tour of Japan in 1980.

An honour that eluded West was the Regius Chair at Oxford when Lloyd-Jones retired in 1989. The burden of administration and advocacy that the job increasingly involved, however, would not have been to his liking and would have stolen time better spent on other things. (He once headed the section on administration in his CV 'Du temps perdu à la recherche'.) The election to All Souls in 1991 was a salvation both for him and the world of scholarship. He was not a natural tutor—that taciturnity made tutorials challenging for undergraduates—and the list of his research students is not long.[62] But he gave generously of his time to those who asked; an untold number of books acknowledge his help. He inspired affection and loyalty among those who knew him well, as his fine Festschrift *Hesperos* attests. Numerous publications were dedicated to him after his death and Balliol College has instituted an annual lecture in his memory.

West died unexpectedly of a heart attack on 13 July 2015. He is survived by his sister Jennifer Lesley Bywaters (born 1947), by his widow Stephanie Roberta West née Pickard (herself a distinguished classicist and Fellow of the British Academy; they met in Fraenkel's seminar and were married in 1960), and by his children

[60] Classicist or ancient historians in the Order before West were Richard Claverhouse Jebb (1905), Henry Jackson (1908), J. G. Frazer (1925), J. W. Mackail (1935), Gilbert Murray (1941) and Ronald Syme (1976). A. E. Housman declined the offer in 1929.

[61] These reports make amusing reading. Letter of 21 February 1991 accepting the Fellowship: 'It may lengthen my life and it will certainly shorten my address.' Upon being made an Honorary Fellow of University College: 'It is a pleasant thought that when I expire the flags will be at half mast on both sides of the High.' June 2003: 'My score of 1 not out was not, as it turned out, a decisive factor in the Fellows' victory' *sc.* in the cricket match against the College staff. I am grateful to the Warden of All Souls College, Professor Sir John Vickers, for sight of these documents and permission to quote from them.

[62] I am aware of Paula da Cunha Corrêa, †Kweku Garbrah, Sophie Gibson, W. B. Henry, J. H. Hordern, †Stephen Instone, Peter Kingsley, Letizia Palladini, N. J. Richardson and Sandra Šćepanović. At All Souls, he was academic advisor to J. L. Lightfoot and P. J. Finglass.

Rachel Ann Dillon (born 1963) and Robert Charles West (born 1965). There are two grandchildren.

At the close of his introduction to *Greek Lyric Poetry*, West wrote the following words, which may provide a suitable envoi:

> It has been an enjoyable task. I do not delude myself that all parts of the end product are likely to give equal pleasure to the reader. But if I have succeeded in opening any eyes, ears, or hearts to some portion of the manifold beauty, wisdom, and wit that shines from these precious remnants of a brilliant culture of long ago, I shall be well content.

He can rest very well content indeed.[63]

Acknowledgements

I am most grateful for assistance of various kinds to Bill Allan, Richard Alston, †Alan Cameron, George Cawkwell, Patrick Finglass, Almut Hintze, Simon Hornblower, Richard Jenkyns, Robert Parker, Christopher Stray and Yuhan Vevaina; to the Warden and Fellows of All Souls College; and above all to Stephanie West.

Note on the author: Robert Fowler is Henry Overton Wills Professor of Greek Emeritus at the University of Bristol. He was elected a Fellow of the British Academy in 2015.

[63] I have made liberal use of published memoirs and obituaries, beginning with those delivered by Jane Lightfoot, Alan Cameron and Robert Parker at the All Souls memorial event, https://www.asc.ox.ac.uk/sites/stage.all-souls.ox.ac.uk/files/Memorials/Martin%20West%20Memorial%20Addresses.pdf (accessed 23 April 2018). Seven obituaries can be consulted via https://www.asc.ox.ac.uk/person /79?m1=6&m2=0 (accessed 23 April 2018): *Oxford Mail*, 16 July 2015 (anon.); *The Telegraph*, 21 July 2015 (staff writers, with input from C. Carey); *The Times*, 24 July 2015 (P. J. Finglass), in Greek at *Ariadne. The Journal of the School of Philosophy of the University of Crete*, 20–1 (2014–15), 294–300; Το Βήμα, 26 July 2015 (A. Rengakos and C. Tsagalis); *The Independent*, 3 August 2015 (G. O. Hutchinson); *The Guardian*, 13 August 2015 (J. L. Lightfoot); *Bulletin of the Council of University Classics Departments*, 44.4 (2015), 1–3 (Armand D'Angour; cf. http://www.armand-dangour.com/2015/08/in-memoriam-martin-west/ [accessed 23 April 2018]). I have also seen H. Bernsdorff, *Frankfurter Allgemeine Zeitung*, 22 July 2015; M. Davies, *The Oxford Magazine*, 2nd week Michaelmas Term 2015; M. Davies, *TW: The Magazine of St John's College Oxford* (2015), 45–6; P. J. Finglass, *Lexis*, 33 (2015), 1–4; A. Fries, *Studia Metrica et Poetica*, 2.2 (2015), 152–8; J. Danielewicz, *Meander*, 70 (2015), 3–7; L. Lehnus, *Rivista di filologia e di istruzione classica*, 144 (2016), 182–98; J. L. Lightfoot, *Gnomon*, 88 (2016), 281–5; eadem, *Proceedings of the American Philosophical Society*, 161 (2017), 285–92; R. Jenkyns, *Balliol College Annual Record* (2016), 27–32.

John Horsley Russell Davis

9 September 1938 – 15 January 2017

elected Fellow of the British Academy 1988

by

PAUL DRESCH

ROY ELLEN
Fellow of the Academy

Biographical Memoirs of Fellows of the British Academy, XVII, 121–143
Posted 7 September 2018. © British Academy 2018.

JOHN DAVIS

John Davis was a key figure in Mediterranean anthropology. He was also a pioneer in applying insights from the literature on exchange to the informal economies of complex societies, and in the use of computers in anthropological research. Educated at Oxford (1958–61) and at the London School of Economics (LSE) (1963 onwards), he spent his most productive years at the University of Kent, after which he returned to Oxford, first as the head of the Institute of Social and Cultural Anthropology (1990–5), and latterly as Warden of All Souls (1995–2008). He died on 15 January 2017.

I

John was born on 9 September 1938, the son of William Russell Davis and Jean (née) Horsley.[1] His father left at an early stage, and John had no clear memory of him; his mother then lived in a *ménage* of independent ladies which lasted through the Second World War; and her second marriage, to a finance director, when John was thirteen, provided him with a readymade pair of older brothers, whom he liked. Though an odd (perhaps even unsettling) childhood, it was not materially deprived, and John was sent first to Dulwich College Preparatory School, then to Christ's Hospital, at neither of which he flourished. At Christ's Hospital he did acquire a love and knowledge of music, and a few inspiring teachers gave him a glimpse of scholarship as fun. Furthermore, he was packed off to Paris for a summer *cours de civilisation*, and in a queue to fill in forms he encountered an interesting English woman much older than he, with whom in due course he had an affair. Back at school, 'I found life even more insufferable than before.'

It somehow seems typical of John that the woman he met in a queue at the Sorbonne turned out to have been the third wife of Bertrand Russell—and, indeed, that John became great friends with her son (later an eminent historian), who was roughly his own age. The pattern of 'contacts' continued. At University College, Oxford, where he won a scholarship to read modern history, John fell in with 'what eventually became the *Private Eye* crowd'. He knew Peter Jay and Margaret Callaghan, and thus something of the worlds of politics and media. Before Oxford, having been more or less asked to leave Christ's Hospital, he spent a year in Italy, but when he began his own work there, years afterwards, the same pattern of interesting friendships emerged. Through an introduction from Paul Stirling, he was taken up by Manlio Rossi-Doria, a professor of economics at Naples who was also a senator in the Italian parliament and a considerable figure in Italian cultural circles.

---

[1] In addition to published work, we have drawn on a number of private sources and personal reminiscences, indicated as quotations but without a citation.

A second-class degree from Oxford, so John said later, put paid to his ambitions as a historian. More important in explaining his move to anthropology was probably an Oxford friend (Lady Russell's son, in fact) pressing on John the interest of what anthropologists wrote. Whatever the case, John became a graduate student in anthropology at LSE, where he encountered Raymond Firth, Maurice Freedman and Lucy Mair. It was Mair, whose sharp mind and writing style he much admired, who supervised the completion of his doctorate (awarded in 1969). Much earlier, he had a letter from Paul Stirling:

> It said in effect that he had no respect for people from Oxford since they thought themselves far too clever; he had no high expectations of anyone who had no training in anthropology; he didn't really want to employ anyone who had only a second-class degree—but he needed someone who could speak Italian to be his research assistant.

Stirling, whom he liked a great deal, was hugely important to John's career. But Ernest Gellner, whom he first heard lecture at the LSE, is spoken of by many who knew John well as no less than a 'father figure', as is Rossi-Doria. Several fellows of All Souls, where John arrived in his fifties, remember him having Gellner to dinner, when John was Professor of Social Anthropology at Oxford and Gellner was the equivalent at Cambridge, and John treating Gellner with marked deference.

John himself later wrote of life as a succession of 'accidents'. Looking back, however, one has a strong impression of someone who, for whatever reasons, constructed a persona early in life (perhaps in his mid-teens) and lived by way of it thereafter. The persona was an established English one, alert to social context in the manner of certain novelists of the period (Ivy Compton-Burnett and Anthony Powell were two of John's own favourites), erudite in a way that was less common in England and with a cosmopolitan, European colouring. From the start, the authorial voice is that of an established figure. For example, here is John, at the age of thirty, on Mediterranean rhetorics of honour:

> They generally have as one of their components the control by men of women's sexuality, and the resulting combination of sex and self-importance makes a unique contribution to the human comedy in life and literature.[2]

Mixed with wide-ranging allusions to history and literature, the assured and measured tone gave his work a feeling of sophistication. Exchange theory, with its technicalities, might encourage dull exposition, but John picks out how 'rationality' in analysts' models filled the place of motive in the simpler kinds of moralising, and thus reproduced a foolish dilemma:

---

[2] J. Davis, 'Honour and politics in Pisticci', *Proceedings of the Royal Anthropological Institute of Great Britain and Ireland*, 1969 (1969), p. 69.

[M]y action was not truly noble because I had an eye to the audience; but if I simultaneously recognise my own duplicity, I am not as other men are—but if I can even think such a thing as that, then I am lost! The dreadful antiphony of regressive self-examination is familiar to us.[3]

An anthropologist need not adopt puzzles that grown-up persons knew in their own lives were unreal. He goes on to say:

It is curious that an absolute division between morality and prestige on the one hand, and self-interest on the other, should have so perpetuated itself in social theory: that what a sensible man brushes away as an adolescent circularity when he is tempted to examine his inner life should be entrenched as the saving grace of a sociological tautology.

II

In 1966 John moved to Kent, like his fellow Italianist Nevill Colclough, following the appointment there in 1965 of Paul Stirling. (At that time academic life still had room to seize upon early talent; completing the doctorate could wait.) They were part of a group from LSE who formed the nucleus of a board of studies, later to become the Department of Sociology and Social Anthropology, and John was on the anthropology staff until 1990. Although Stirling established anthropology at Kent, he himself had been appointed as Professor of Sociology. John was Kent's first Professor of Social Anthropology, a personal chair accorded him in 1982, and he often said that he was 'made at Kent'. This was where his major work was done. Being among mainly sociologists at the beginning surely pushed him towards a broader social anthropology, particularly to examine the British informal economy, but Mediterranean anthropology was also in full bloom. J. G. Peristiany's conferences in Athens were a highlight, Kent became a central place and Mediterranean anthropology formed a kind of travelling house party. As Michael Gilsenan puts it:

I don't think that I have ever quite recaptured the atmosphere and stimulation of those conversations, meetings and conferences in Kent, Rome, Zaragoza, Galicia, so many places. Ernest Gellner added enormously to the engagements and felt a real link with Kent, Sydel Silverman and Eric Wolf became friends. Provocation, argument, and lots of food and drink, those were the rules.[4]

The period, although one of UK university expansion, was one in which many anthropologists presented themselves as 'social scientists' in order to secure positions

[3] J. Davis, 'Forms and norms: the economy of social relations', *Man* n.s. 8 (1973), 162.
[4] M. Gilsenan, Memorial Address, All Souls College, Oxford, 24 June 2017, p. 10.

as well as to engage with new intellectual trends. In the early days at Kent, John both benefited from and contributed to the intellectual synergy between his work and that of colleagues in sociology, such as Ray Pahl, the political sociologist Krishan Kumar, Derek Allcorn (whose theoretical acumen and sense of humour he much admired) and Frank Parkin. It was for Parkin that, in 1974, John persuaded Gellner to hold the press deadline of the *European Journal of Sociology* to accommodate a *faux-Marxisant* analysis of Beatrix Potter, in which it was claimed there could be no such thing as 'an innocent reading'.[5] The Kent years were punctuated with sabbaticals at the Middle East Technical University in Ankara (1971), the Free University of Amsterdam (1977–8), the University of California, Berkeley (1980) and the University of Aix-en-Provence (1981). In Amsterdam John met Dymphna Hermans, whom he married while at Berkeley. Dymphna, herself a social anthropologist, later undertook field-work in Cambrils, a village on the Costa Dorada of Catalonia. They were different but complementary characters: he with controlled and conventional middle-class manners, she effervescent and outgoing. The couple were to have three children—Michael and Henry (born 1983) and Peter (1985).

Not a technical person in the normal sense, John could treat technical matters with great intensity. This was as evident when he took to bricklaying at his house at Kent as in his long enthusiasm for photography. More importantly, it was John—an otherwise more unlikely figure one could not imagine—who founded the Centre for Social Anthropology and Computing, and for a while placed Kent at the forefront of computing applications that are now standard. Although Marie Corbin and Paul Stirling, as early as 1969, had used a Social Science Research Council (SSRC) grant and the Didcot Atlas Computer Centre to reconstruct family and kinship data from census records, it was John who routinised computer use. In Berkeley he had been much impressed by the work of the Language Behavior Laboratory under Eugene Hammel and Brent Berlin. On returning home, John persuaded the SSRC to sponsor a workshop at Kent in December 1983 on 'Computing and anthropological research'. Over the following years he introduced applications that we now take for granted (bibliographic and other databases, text production and statistical packages) as well as specialist applications for handling kinship. A *Bulletin of Information on Computing in Anthropology*, initially edited by John, ran from 1984 until 1992. Characteristically, he christened the Kent anthropology server 'Lucy', not to memorialise the Ethiopian fossil hominid (as many thought), or the linguistically capable chimpanzee (as some thought), but to honour Lucy Mair.

As a teacher, John is remembered for 'Understanding other cultures' (a joint course with philosophy), a history–anthropology bridging course and—perhaps with

---

[5] R. and C. Parkin, 'Peter Rabbit and the *Grundriße*', *European Journal of Sociology*, 15 (1974), 181–3.

more ambivalence—for *L'Année sociologique*. This latter attempt to bring rigour to the study of Durkheim's inheritance was short-lived, not least because a condition of registration was fluency in reading French. More popular was his creation of 'Potlatch', a teaching game that sought to capture the dynamic properties of the eponymous Kwakiutl institution of competitive exchange. Entertaining and amusing both as guest and host, during the 1980s he would invite the Tuesday anthropology research seminar back to his spacious kitchen in St Thomas Hill (John was always a fine cook), where discussion would often continue over food and drink, even after John himself had withdrawn to bed. For Krishan Kumar, there was in all this a 'generosity of enthusiasm'.[6] John had a particular reputation for generosity towards the young although, sadly, he had few research students of his own who might have served as torch-bearers of his reputation. His writing had to speak for itself.

III

John's anthropological work in Italy began in 1963 with a six-month stint as Paul Stirling's research assistant, at a time when the Mediterranean drew wide interest. Such authors as Julio Caro Baroja had written well on Spain, but it was Julian Pitt-Rivers's *People of the Sierra* (London, 1954) that forced English-speaking anthropologists to pay attention. Stirling had worked in Turkey in the late 1940s, although he was slow to publish (his *Turkish Village*—London, 1965—would still have been in draft when John was his assistant). And John Campbell, who studied under Peristiany, was beginning to publish on Greece as John started work in Italy. (All of these people had been co-opted into British anthropology by E. E. Evans-Pritchard at Oxford.) Pitt-Rivers had edited *Mediterranean Countrymen* (Paris, 1963), and a series of important collections edited by Peristiany began in 1965 with *Honour and Shame* (London).

The first fruit of John's own efforts was a piece on the card game Passatella, published, it seems, with Maurice Freedman's help. This derived from a sustained commitment to a wine shop in Bolgiaquinta and two-and-half months of playing sociable cards. Confusingly for the newcomer, no doubt, the face values of the cards were not their scoring values; nor would it have been plain to start with what the game's attraction was. Despite elaborate shows of secrecy and deliberation, there was little room for calculation in the card play itself, since the deck was taken in and reshuffled after two inconclusive rounds of discards. The fun began when two 'masters' emerged by winning a hand. They then apportioned the drinks, a lengthy and complex

---

[6] Gilsenan, Memorial Address, p. 10.

process that *did* require strategy, 'around the cards went again, and (largely tacit) alliances were formed and reformed in the process of again allotting drinks, with people co-opted, betrayed or abandoned'. The article suggests an already fine grasp of Italian, a sharp eye for detail and a knack for sociability. Not everyone could have joined the wine shop regulars. A circle of players (all men, of course) were 'friends', some had been allies at the Passatella table for a quarter of a century, and the game exemplified, in a friendly way, the same concerns with patronage, rivalry and dishonesty as informed real life in a place where men were entangled with each other in the nature of things and resources were scarce. A loser was *fesso* (gullible, stupid), a winner was *furbo* (cunning) and admired as such.

The article exhibits several features that would characterise John's later work. The writing is exact and meticulous. Grand theories of anthropology are conspicuously not appealed to, and a novel (in this case Roger Vailland's *La loi*: Paris, 1957) may be a better source than 'social science' writing. The analysis is very much John's own and he sets it up with care: practical action usually has an overall end in view, but a formal game involves actions, 'each of which has its own moral value and excitement', while the fact that a game may have a goal 'does not help us to describe the game'.[7] John, meanwhile, must have exhibited a good deal of *furbezza* himself, not least to have taken notes to judge the detailed forms of reciprocity while presumably counting his own cards and chaffing his neighbours at the table.

By the time the piece saw print, John was engaged in fieldwork on his own behalf, in a town of about 15,000 people on the arch of the 'boot' of Italy. This was the subject of his PhD thesis (1969), of articles and of a monograph, *Land and Family in Pisticci* (London, 1973). In the same year as he gained his doctorate, John won the Royal Anthropological Institute's Curl Prize for an essay on 'Honour and politics in Pisticci'.[8] John begins by pointing out that the English term 'honour' does not denote the same thing everywhere. He is keen, as in the piece on card games, that we not mislead ourselves by asking what honour is 'for'. He expresses, also, a distrust of 'social structure' being invoked by anthropologists to recover a presumed coherence of belief where in fact there is none. But his own account stresses ranking and inequality, and he attaches the fact that the prestige of a person or a family is based on a moral unity, despite the many different criteria used to judge their actions, pertaining to control of resources. The theme would recur in his later work. For the moment, the relation of ideas to behaviour, as he chose to put it, was left hanging. On the one hand there were individual persons, on the other was something very like economics. This, of course, trespassed on broader questions.

---

[7] J. Davis, 'Passatella: an economic game', *British Journal of Sociology*, 15 (1964), 192.
[8] Davis, 'Honour and politics'.

In *The Moral Basis of a Backward Society* (Chicago, 1958), Edward C. Banfield, an American political scientist who went on to be a presidential advisor, had blamed southern Italy's problems on a (non-economic, even anti-economic) 'family-centred ethos'. John, with a number of scathing remarks on Banfield's parochialism, insisted this was no kind of explanation. Family and community were inseparable, social control deserved attention and people's choices made sense when one realised how constrained was the range of practical options. Choice was one part of John's polemic. The other was economics. Unless someone has the power to block the process, says John didactically, 'Moral ideas always give way to economic opportunity.'[9]

The monograph *Land and Family* shows his usual virtues. Again, anthropologists appear if they are useful, not otherwise (there are five obvious names in the bibliography, most of them receiving just a nod). At the end of the book John rejects the idea that all Pisticcesi life might be seen as 'networks' and 'negotiation'. What he had in mind were such works as Fred Bailey's *Stratagems and Spoils* (Toronto, 1969). His own concern is

> to know why particular people are face-to-face at all; what is it that brings *them* together? Why do *those* two or three have need of each other? Why is it that discussions between families about marriage endowments are concluded more equably than discussions about the same endowments within families?[10]

His answer is 'social structure', but that in turn is a product of Italian history, and Italian thinkers, novelists, historians and sociologists are more prominent in John's bibliography than are anglophone anthropologists. (A slightly later piece in a Festschrift he organised for Lucy Mair shows remarkable confidence with language and dialect, albeit Nevill and Ripalta Colclough helped transcribe the tape recording from which John worked.[11]) His doubts about what he knows for sure and what he infers less surely are clearly marked, and his arguments are laid out in carefully constructed prose. This is not to say the writing makes concessions. Indeed, John was not above flaunting some gratuitous learning. The word 'paratherns' crops up with no explanation and little context, when even the more common Latinate *parapherna* might more easily have put one in mind of the Greek *paraphernalia* and thus of goods that go with the bride. Presumably, many readers turned to the dictionary.

Here was a place, Pisticci, where nearly everyone owned a piece of land, and most had interests in other pieces, but few people owned enough to get by; where underemployment was chronic and patronage was rife; and where neighbourhood, shared

---

[9] J. Davis, 'Morals and backwardness', *Comparative Studies in Society and History*, 12 (1970), 351.
[10] J. Davis, *Land and Family in Pisticci* (London, 1973), p. 162.
[11] J. Davis, 'How they hid the red flag in Pisticci in 1923, and how it was betrayed', in J. Davis (ed.), *Choice and Change: Essays in Honour of Lucy Mair* (London, 1974), pp. 44–68.

descent and intermarriage made life intensely sociable but no one could fully trust their friends. Why was it like this? John placed enormous weight not only on his own surveys but on local government archives, combing them as a historian would but anthropologists at that time rarely thought to do, and found as a baseline a cadastral survey of 1814:

> We cannot say, I think, that economic relations are basic and that kinship is a mere descriptive idiom emanating from them … What we can say is that certain features of the present-day kinship system … are new, not more than 150 years old. They appear also to be coeval with the distribution of land and the diversification of the local economy.[12]

As a perceptive reviewer of the book chose to phrase this, 'poverty is the South's way of being modern'.[13]

The book is of its time. Although migrant labour is mentioned at several points, almost as much space goes to labourers coming in from nearby Lecce to share-crop tobacco as to those Pisticcesi (we are not told how many there were) who worked elsewhere in Italy or in Germany or North America. 'Globalisation' was not yet a word to conjure with. Nor, conspicuously, is class discussed. John's position seems always to have been that unless there is explicit class consciousness locally, class is of no analytical use (the national salience of an Italian Communist Party did little to convince him). The one citation of Marx is thus from 'The Eighteenth Brumaire'—the line on peasants as a disorganised mass, like potatoes in a sack. Ranking by honour, says John, meant Pisticci was not like that: 'It is difficult, moreover, to imagine a continual class consciousness in a society in which rather more than half the population has some rights, however small, in the principal form of capital.'[14] *Land and Family* was not widely reviewed. It was, however, widely cited, and still is. An Italian translation came out in 1989 and the English version was reissued in 2004. The message to anthropologists remains what it was. Although one could, if one wished, make south Italian peasant society seem 'exotic', the beginning of wisdom about a place like Pisticci was to recognise that 'the Italian normal, the local idiosyncratic, the European commonplace are there combined'.[15]

---

[12] Davis, *Land and Family in Pisticci*, p. 160.
[13] J. W. Cole, 'On the origins and organization of South Italian poverty', *Reviews in Anthropology*, 2 (1975), 84.
[14] Davis, 'Honour and politics', p. 78.
[15] Davis, *Land and Family in Pisticci*, p. 1.

IV

At the time when *Land and Family* appeared, Jane and Peter Schneider were at work in Sicily, John and Marie Corbin were in Andalusia, Eric Wolf and John Cole were investigating an Alpine valley, and Wolf and Bill Schorger had for years run conferences on the Mediterranean, the last of which convened at Kent. American authors were bringing to bear fresh intellectual resources. Two of Peristiany's collective volumes on Mediterranean themes were in print,[16] and Braudel's great work was becoming available in English,[17] although one would guess with confidence that John had read the much earlier French original. His *People of the Mediterranean* (London, 1977) summed up where he thought anthropology stood. Perhaps surprisingly, given his own intellectual background, John, with few exceptions, left aside historians and others to concentrate on (mainly English-speaking) anthropologists. The book, which was widely cited, and in 1983 translated into Spanish, was described as 'a milestone that marks the coming of age of a Mediterranean social anthropology',[18] which was surely the author's aim. If he was a generation too late to be a pioneer, he was certainly intent on putting the older generation right, and he describes Mediterranean anthropology, not altogether fairly, as 'an almost complete museum of pre-modern research techniques'.[19]

Again, John could not resist an occasional obscure word (for example 'erogate'), but as ever his prose is in general a model of clarity. The controversies that arose were therefore not about ambiguity. Some readers doubted whether a discrete 'anthropology of the Mediterranean' was a wise idea, others whether John's view of the subject as the sum of existing ethnographies was viable. Certainly the book pulled together a vast amount of anthropology, and it was characterised, years later, by a colleague in the French system, as 'a solitary and promethean plea for a more resolutely comparative and historical approach'.[20]

Most studies to date had been highly localised, as anthropological studies often are, and setting one such study beside another had its limitations. By this time an Austrian colleague could report that half the population (perhaps she meant half the male population) of the village in Greece that she wrote about worked in or around

---

[16] J. G. Peristiany (ed.), *Honour and Shame: the Values of Mediterranean Society* (London, 1965); J. G. Peristiany (ed.), *Contributions to Mediterranean Sociology* (The Hague, 1968).
[17] F. Braudel, *The Mediterranean and the Mediterranean World in the Age of Philip II* (London, 1972).
[18] J. Boissevain, 'Towards a social anthropology of the Mediterranean', *Current Anthropology*, 20 (1979), 81.
[19] J. Davis, *People of the Mediterranean: an Essay in Comparative Anthropology* (London, 1977), p. 2.
[20] D. Albera, 'Anthropology of the Mediterranean: between crisis and renewal', *History and Anthropology*, 17 (2006), 126.

Stuttgart. Nor was timescale less problematic than geographical range. John himself had shown how, in Pisticci, patterns of marriage and land tenure were not very old. Behind such changes there seemed nonetheless to be a Mediterranean *longue durée*, and resemblances between the north and south shores of the sea were often striking; the record was full of 'institutions, customs and practices which result from the conversation and commerce of thousands of years'.[21] Without history in a more detailed sense, however, anthropology ends either with vague talk of 'culture areas' or with implausible appeals to common causes such as climate. It was not until 2000 that Peregrine Horden and Nicholas Purcell (a medievalist and a classicist respectively) suggested a way around the impasse by stressing ecological variation on a local scale, 'connectivity' among localities and the production of difference through exchange.

People of the Mediterranean is combative. The shortcomings of existing works receive more emphasis than their strengths, and the tone is often acerbic, with a monograph on one Italian village tagged as 'the second worst book on Mediterranean anthropology' and the reader left to guess which book was the worst (presumably it was Banfield's).[22] A wife and husband team are singled out for 'the amount of ingenuity they have expended to make their data irrecoverable'.[23] Most memorably, Julian Pitt-Rivers's early classic on Andalusia is described as 'a tangle such as only a pioneer's licence could justify'.[24] Among Pitt-Rivers's papers, it seems, is a long and very angry response that he never published, but his letter on 'The value of the evidence', in *Man* (1978), was sharp enough.[25] John, for his part, might have phrased his initial comments differently. His reply to Pitt-Rivers, also in *Man*, retrieved certain points that mattered.[26] They disagreed over the propriety of publishing details of a kind that make people and field sites identifiable, where John's inclination was towards transparency. John also insisted that, while equality may be an aspiration, as it was in Andalusia, honour is hierarchical. His position was that which he had argued in his 1969 Curl Prize essay, and 'egalitarian institutions' are in his view a rarity; power goes with wealth, honour reflects this and anthropologists, he thinks, deceive themselves by analysing 'values' in isolation. His central point, which this echoes, concerns how to compare cases and thus what the bases of comparison might be. Somewhere, he feels, must be 'crude material differences in wealth' that would allow one to set cases in order.

---

[21] Davis, *People of the Mediterranean*, p. 13.
[22] Ibid., p. 73.
[23] Ibid., p. 81.
[24] Ibid., p. 93.
[25] J. Pitt-Rivers, 'The value of the evidence', *Man* n.s. 13 (1978), 319–22.
[26] J. Davis, 'The value of the evidence', *Man* n.s. 13 (1978), 471–3.

Years later, in an unpublished note, he gave vent to his view of Pitt-Rivers, though how their mutual dislike first developed remains obscure. Oxford anthropology in the 1950s, says John, had depicted other cultures as implausibly symmetrical, patterned and coherent. 'The most exquisite of the snowflakes was Julian Pitt-Rivers, an accomplished solipsist.' 'Every inch the Old Etonian gent', he had opposed John's views on honour, so John supposed, on the grounds that 'it was vulgar and middle class to consider the distribution of wealth; what a truly sensitive anthro[pologist] could perceive was the "essence"—egalitarianism'. John's own rejection of class as an explanatory principle, one might think, befits successful *novi homines* for whom class relations are merely something to clamber about among. Behind the squabble, however, lay more profound questions:

> It is clear that when Pitt-Rivers has a problem ... he will cast around, invoking Voltaire, Haro, Lope de Vega, Odysseus, and a romantic novelist called La Picara Justina. The point is this: there are assumptions of continuity—historical and geographical— which are not spelled out, and which should be argued, if an impression of potpourri is to be avoided.[27]

John wanted causal explanation. The categories seemed to him fairly unproblematic. Pitt-Rivers was interested in categories and if some of these, such as hospitality, recur in one form or another everywhere, then Voltaire and Homer might indeed belong on the same reading list.

People of the Mediterranean rejects 'explications of the conceptual intricacies of ambiguous notions' as anthropology's true aim.[28] John himself, however, was intrigued by ambiguity, and a succession of pieces attempts to resolve the puzzle he had earlier set himself by opposing ideas and behaviour. Besides articles on 'Forms and norms' and on exchange theory, one thinks of his later claim that all rules have parasitic rules explaining them away: 'it is inconceivable to us that a society that "has" rationality should not also "have" sophistry'.[29] Later still, in a piece on irrationality, some of the colour professionally excluded from the Pisticci monograph was placed on display, as when a petty trader was recounted as being beaten up after a 'friend' let slip that the trader was having sex with an underage girl. The trader escaped through sophistry:

> You seem to be losing an argument for one proposition (it is right that merchants should have sexual access to the simple minor daughters of their social inferiors). If you are barefaced enough ... you can try to shift the argument to another where you have a better chance (it is wrong to betray a friend). When you have won that argument,

[27] Davis, *People of the Mediterranean*, p. 253.
[28] Ibid., p. 93.
[29] J. Bousfield and J. Davis, 'What if sophistry is universal?', *Current Anthropology*, 30 (1989), 518.

you can boldly claim that you have in fact won the other.[30]

This manoeuvre was no different, said John, from that of social scientists, who recuperate in their writing all manner of inconsistency and should instead give stupidity its due by creating 'ethnographies of doubt and argument'. John offered this to Gellner as 'a small bouquet of absurdities'. A contrast between the coarseness of the case and the subtlety of the presentation may have appealed to both men, but the rationality of 'economics' and the like as a unitary explanation of social life was certainly demoted.

V

Paul Stirling, who had drawn John to Kent, had worked in Turkey. Pierre Bourdieu's writing on Algeria had come to English speakers' notice through Peristiany's *Honour and Shame*. Evans-Pritchard and then Emrys Peters had long ago written on eastern Libya, and in the pre-war period French ethnographers had done excellent work in North Africa, which John had begun to explore in his Mediterranean volume. More recently Gellner and, less successfully, Clifford Geertz had published on Morocco. A professional Mediterraneanist, which is plainly how John saw himself, should ideally work on both shores, and as early as 1973, it seems, he set about learning Arabic.

In 1975 he began work in Libya, which went on, at intervals, until 1979. (At the end of his first stay John had his field notes stolen, amongst other possessions, from the back of a Land Rover in Rome; he searched unsuccessfully through Rome's rubbish heaps for days.) Libya was not an easy place to work, and one suspects good Italian contacts played a part in getting started. The literature was scanty, and no one with careerist ambitions of the simplest kind, and thus a need for fieldwork one could complete on schedule, would have aimed for Kufra in the deep south. Intermittently, a war was going on that involved Chad. Across tribal and linguistic divisions (Mgharaba and Zuwaya, Tubu and Arabic) that had nothing to do with national boundaries, fighters and supplies moved back and forth, not least through Kufra where, it might reasonably be assumed, there were many things best not noticed: 'Watch the wall my darlings'. These days one would be at hazard of killer drones.

The questions of transparency on which John had disagreed with Pitt-Rivers looked different in Libya, and John was careful not to compromise his sources. But the same knack for sociability that allowed the early paper on Italian card playing clearly worked in Ajdabiya and Kufra. John got on well with Zuwaya traders. And the local security official who, presumably, was meant to keep an eye on him, became a

---

[30] J. Davis, 'Irrationality in social life', in J. Hall and I. Jarvie (eds.), *The Social Philosophy of Ernest Gellner* (Amsterdam, 1996), p. 449.

friend who in the end, we suspect, got him access to a hoard of legal documents that John duly photographed. *Libyan Politics* (London, 1987) mentions illness; it refers also to the many hours each day given over to field notes. The funny stories that inform most fieldwork are not in print, although one of them appears in a later reminiscence. Having clambered on top of a car to remove a faulty light fitting on the gateposts of a local school, the anthropologist duly checked with the locals that the current was switched off, inserted a pair of pliers to remove the broken part and was blown off the roof of the car by a whopping electric shock. Very cross, he withdrew to his tent, where the schoolmaster's brother came to arrange a reconciliation, if need be by sacrificing a sheep. John duly went to see the schoolmaster and forgave him, as one is meant to do. It was all rather silly, if mildly revealing, 'And in any case, what would you do with a dead sheep?'

The book 'is at once very "off-beat" and very good indeed'.[31] The scrupulous surveys John conducted of kinship and residence patterns in small Libyan towns were conventional, but the dominant theme was the nature of the Libyan state, and here he felt free to explore fresh ideas. There were 'cadres', certainly, but

> in Ajdabiya revolutionary friend and revolutionary foe were embedded in the same structures of descent and marriage and neighbourliness. And this was so for every cadre in Ajdabiya and Kufra and the other smaller oases, and it applied to the members of other tribes as well as the Zuwaya: the cadres of the revolution were members of their society, and an important part of that society was organized in lineage and tribe.[32]

There was no simple distinction between state and society, therefore, although the ruling circle monopolised decisions on major topics, and *hukuma 'arabiya*, 'Arab government', here really meaning 'people's government', was a common phrase to discuss a world where, ideally, there would be no institutions of top-down authority. Qaddafi's government had seemingly boundless oil wealth and no need to organise the population for productive ends; the population could draw on government stipends as of right, and did so. *Libyan Politics* explores this with the aim of extending readers' imagination of political life. The book was widely and favourably reviewed. A French translation came out in 1990. It provided, as one reviewer said, 'a welcome relief from the ethnocentric and politically motivated mystifications of most Western writing concerning modern Libya'.[33] It was also something of an escape from the narrow terms that had so far bound John's anthropology.

[31] D. Hart, review in *International Journal of Middle East Studies*, 21 (1988), 433.
[32] J. Davis, *Libyan Politics: Tribe and Revolution* (London, 1987), p. 93.
[33] R. Fernea, Review in *American Anthropologist*, 90 (1988), 1013.

VI

'Exchange' runs through John's work from first to last. This is not surprising given the early influence of Raymond Firth, but whereas Firth maintained that formal economics could serve in studies of non-monetary economies, John explored the limits in everyday life of economistic language. 'Gifts in the UK economy' drew attention to the volume of transactions, produced and acquired in the formal economy, that 'disappeared' into a sub-economy of gift where different rules applied;[34] 'Forms and norms' used the example of Tupperware parties to show how firms were able to harness these sub-economies to capitalist enterprise, but criticised analysts who would 'see the everyday inanities of love analysed with the aid of Paretan optima and indifference curves'.[35] Equally, he was sceptical of the stark distinction between gift and commodity that anthropologists such as Chris Gregory and James Carrier had made. In 'The particular theory of exchange', John develops a commentary on exchange theory of the Peter Blau variety, which is where he comes closest to a full-blown critique of economic formalism.[36] His later volume on *Exchange* (Minneapolis, 1992) sets out in fewer than 100 pages an approach to the anthropology of exchange, and his work on the informal economy of Western societies had an influence beyond anthropology, for example in Ray Pahl's *Divisions of Labour* (Oxford, 1984). It is surprising, however, how little John draws on his Italian and Libyan experience. With a second, and later, intellectual thread, which was history, the influence of fieldwork is more prominent.

One of the oddities of John's biography is that, despite his undergraduate background in history and his dislike of 'timeless' ethnography, he expressed no broad views, even in conversation, of historical method or of well-known historians. Analytically he had gone no further than to point out the shortcomings of Oxford anthropology's vague talk of history in the years around 1950. In the Mediterranean book, history figured largely as a set of straightforward 'causes'. The exception was Carmelo Lisón Tolosana's description of a Spanish village, where the younger generation rejected the older generation's view of the past to establish their own view of a possible future. This sets up John's own concluding section: 'Covertly, it might be said, Lisón Tolosana produces an account of how people make history and consume it. It is "living history" in the sense that it is incorporated in an analytical way into the account of a changing social structure.'[37] A review article in 1980 considered how

---

[34] J. Davis, 'Gifts and the UK economy', *Man* n.s. 7 (1972), 408–29.
[35] Davis, 'Forms and norms', 163.
[36] J. Davis, 'The particular theory of exchange', *European Journal of Sociology*, 16 (1973), 151–68.
[37] Davis, *People of the Mediterranean*, p. 248.

American authors whom John admired 'consumed' history in writing on Sicily, an Alpine valley and a town in the Piedmont.[38] John's Libyan work addressed how history was 'produced' by Zuwaya tribesmen, whose

> relations were for the most part those of patrilineal descent. The framework of analysis was not narrative nor progressive nor dialectical but genealogical. And consequently their history consisted of events and lives which carried moral and political precept: of incidents loaded with an interpretation.[39]

In 1983, a piece on 'History in the making' began to generalise Lisón Tolosana's insight,[40] and some years later John published a wide-ranging comparative chapter in a volume that, ironically enough, was informed by Oxford ideas, though of a later vintage than those that John had often condemned. 'The social relations of the production of history' became widely known.[41] Briefly, Lisón Tolosona's generational history and Zuwaya lineage or segmentary history were both contrasted with Qaddafi's newly national history, whose cumulative narrative was part of the modern package in which every nation-state has a history just as each has a flag and postage stamps.

Easily missed, but striking once noticed, are recurrent references in John's work on history and exchange alike to what he now dubbed 'state's men', those who drive informal transactions into the taxable economy and memory into national history. John's interest in such figures increased, perhaps for obvious reasons. In 1980, soon after he completed his Libyan fieldwork, the Public Accounts Committee of the House of Commons had it thrust before them as an example of waste, although, by his own estimate, he had spent on top of his usual salary only £11,000 or £12,000. This was not the generally tolerant Britain in which John had begun his anthropology. The Jarratt report in 1985 urged universities to define their objectives, and funding of many sorts was disappearing; by 1993 John could mention in passing a resemblance between the views of Lady Thatcher and those of Colonel Qaddafi.[42] By the end of that decade, and throughout the next, British readers of *Libyan Politics* must often have shared John's appreciation that politicians who invoke the popular will as legitimising their schemes soon find themselves adrift and lonely:

> Hence the spectacle of rulers putting their ears closer and closer to the ground in the hope that they would eventually hear some whisper that they could recognize as the

---

[38] J. Davis, 'Anthropology and the consumption of history', *Theory and Society*, 9 (1980), 519–37.

[39] Davis, *Libyan Politics*, p. 206.

[40] J. Davis, 'History in the making', in H. Nixdorff and T. Hauschild (eds.), *Europäische Ethnologie* (Berlin, 1983), pp. 291–8.

[41] J. Davis, 'The social relations of the production of history', in E. Tonkin, M. MacDonald and M. Chapman (eds.), *History and Ethnicity* (London, 1989), pp. 104–20.

[42] J. Davis, 'Social creativity', *Anales de la fundación Joaquín Costa*, 10 (1993), 258–9.

authentic voice of the people. They did not and, feeling themselves under threat, became increasingly autocratic and repressive.[43]

An easy thing to set about repressing in autocratic style, with higher aims in view, was academic autonomy. The criteria of success were therefore shifting.

In an unpublished note, written in later life, John says much about the virtues of Paul Stirling: 'And yet he never made it to the top: he was not FBA; he was not an officer of the ASA or RAI, not of any international academic organization, apart … from an Anglo-Turkish association of one kind or another.' This 'getting to the top' plainly interested John more than it had Stirling. John was thus delighted by his election to the British Academy in 1988, and one might have thought that, if the matter troubled him, he could have defined as 'the top' wherever it was that he was pursuing his own interests. In the newly hostile world of 'institutional reviews' he had suitable initiatives in hand. Apart from his involvement in computing, the book on exchange was all but finished, there was work subcontracted but overseen by John on his Libyan documents, a more general review was in progress on Mediterranean marriage and divorce, and a supervised project on Arab marriage in Britain. He had friends and valued colleagues in Kent, a home and a full social life. Yet in 1989 John applied to Oxford. The state of UK academic life may have played its part in this decision.

VII

The Professorship in Social Anthropology at Oxford is attached to All Souls College. When John was appointed, there was predictable discontent from Pitt-Rivers. Anthropologists teaching at Oxford were not upset, however, and when John arrived as professor, in 1990, he did so in a civilised way. His predecessor, Rodney Needham, had given an inaugural lecture marked not only by uncontrolled emotion about his own wartime military service but by unfortunate denunciations of colleagues. John, by contrast, got it right. 'Times and identities' found good in his new colleagues, explained what social anthropology could do, and suggested to the university some attractive ways to consider society, history and ethnography.[44] He gave the subject a certain presence.

John's sociability was much in evidence, as always, but the beginnings were slow. Invited to co-chair the weekly departmental seminar, he said hardly a word all term, as if gauging what he might be dealing with. Wisely, he leant heavily on Peter Rivière, 'an equable man', as John remarks. Occasionally, it must be said, John got things

[43] Davis, *Libyan Politics*, pp. 135–6*b*.
[44] J. Davis, *Times and Identities: an Inaugural Lecture Delivered Before the University of Oxford on 1 May 1991* (Oxford, 1991).

wrong. His social relations with women were sometimes clumsy; his competitive assessments of certain Oxford anthropologists as cleverer than others got back to the objects of his judgement more quickly than he realised and seemed odd to people less interested in precedence. Behind such faux pas, one gathers, was a deep unease. A friend of his speaks more generally of 'insidious whispers' of doubt, not always successfully overcome.[45]

As colleague and professor, John was a genial and supportive presence, with some admitted oddities. His computing interests, for example, took the form of an obsession with introducing specific servers and networks that left those who used PCs all but cut off from the world. In general, however, he worked to include people. At seminars convened by himself, not at those where he felt more vulnerably on display, he drew together anthropologists, historians and others, and set a tone of sophisticated curiosity. Administratively, however, the university was a constant frustration. Committees overlapped with committees, none with the power to decide much, and the central administration produced memoranda that were long, badly organised and poorly written. At least one of these John took the time to rewrite and send back with a note explaining what such a document should look like. Back came the counter-reply that whoever it was could not be blamed, and their mother was sick. John's patience at last ran out. Found by a colleague one day slumped in his office armchair, with the inevitable empty coffee mug and the ashtray with his pipe on the floor beside him, surrounded by piles of paperwork, John said in quiet despair, 'I think I have had enough of this. I shall look for another job.' But this was John; the job that presented itself was thus the Wardenship of All Souls.

John by now had some years' experience of the college, yet he claimed that his election (in 1995) came as a surprise to him. It certainly did to others. It was less of a surprise, it seems, to people within All Souls, where, as a colleague remembers, 'he won the right, the centre-left, and the "nice people" vote', to come out ahead in the first round. A key component of support came from younger fellows. Part of John's appeal among those older was, apparently, the contrast with his predecessor in anthropology: the subject has a certain attraction for those not within it (a touch of the exotic, perhaps; a hint of adventure in remote places), and here was an anthropologist who was urbane and friendly.

The job, of course, is an odd one, not least because it represents the public face of an esoteric institution that attracts much speculation. Wardens may attend formal occasions outside the university by virtue of their office. Locally they are drawn into such matters as the High Street in Oxford being choked by buses. Beyond this they are likely drawn into claims that private papers should be open to public scrutiny, which

---

[45] Gilsenan, Memorial Address, p. 4.

(whatever his position against Pitt-Rivers may have been some years before) John was not going to have happen with respect to the long-ago appointment of Isaiah Berlin. 'We don't let people see current and active files … Sir Isaiah's file doesn't just concern him but also concerns college business and other people's business.'[46] Even before he took up the post, John was accosted by a journalist, who reflected on a wasted afternoon by dubbing him 'the mysterious Professor John John'.[47] His anthropological colleagues hardly thought him so. Indeed, at least one of them found him puzzling for quite the contrary reason: 'The odd thing about John is, What you see is what you get.'

What one saw depended on where one stood, but there was a transparent innocence in some of John's concerns. For a brief time, for instance, his professorship at Oxford had coincided with the Cambridge professorship of Marilyn Strathern, and John would refer to her on occasion as 'Sister Cambridge' and himself thus as 'Brother Oxford'. In print John was respectful of Strathern's work. Yet in private he often related a story of his predecessor as Warden (Sir Patrick Neill, later Baron Neill of Bladon) being urged by Mary Douglas to support Strathern, presumably for the Oxford professorship. Neill had said that he could not understand what Strathern wrote. 'Well,' said Douglas, 'anthropology of course has its special views and methods that non-anthropologists might not grasp.' 'That is not the problem', said Neill, at least by John's account: 'It's her *prose* that I can't understand.'

If All Souls itself is less exotic than journalists hope, it does have peculiarities. The 'mallard song' is sung twice a year, and in every hundredth year a procession, with flaming torches, 'hunts' a mythically gigantic mallard. Such a year was 2001, and, under John's aegis, a verse was restored that refers to the creature's 'swapping tool of generation', though nobody knows why the genitals of male ducks should matter. That of course made the newspapers. It all seems a long way from Pisticci or the deserts of Libya, but John was entranced. An Oxford outfitters at one time had on its racks a surprising number of ties with duck motifs. Were these because of interest from a particular customer? 'Tubby chap, glasses, bright red braces?' 'Yes, sir. Do you know the gentleman?'

This commitment to College had a price. If the Warden is required to live in College, there is no requirement that he live there seven days a week or that his home be permanently open, but John's hospitality meant his evenings at high table were complemented by meals he cooked himself in the Warden's lodgings, and the line between College life and family life collapsed. Dymphna left to live in Wytham, and she and John divorced in 2006.

None of this meant an end to John's wider professional life. He was President of

---

[46] 'Closed files on open minds', *Times Higher Education Supplement*, 9 May 1997.
[47] V. Grove, 'The warden who would not bare his soul', *The Times*, 24 June 1994, 15.

the Royal Anthropological Institute from 1997 to 2001, and his presidential address, in 1998, was very much in character, surveying, with a detachment that not all academics could feel, the position of British universities. Berlioz's encounter with the Académie des Beaux-Arts provides a case of the creative mind encountering mindless rules. Romanticism, however, is of little help to us, says John, and we are compromised before we start:

> Many academics participate in the bureaucracies that control their lives: we sit on the boards of the councils. We may be champions of academic freedom, but we are not all champions *à l'outrance*; we are drawn into complicity in the national councils in the hope that we will be able to defend our disciplines, in the fear that if we don't join someone else, less admirable and staunch than ourselves, will be invited to replace us.[48]

The everyday form of resistance can only be 'irony, tempered with as much compassion as we can muster'. While politicians may pursue a fantasy of 'business methods', collegiality has its own institutional history that spans seven centuries, a view more easily taken, perhaps, if one is part of an ancient college.

John's interests had by now moved on. At some point (we cannot pin down the date) he was heard to say that he had 'lost his faith' in anthropology, an odd thing to say in that anthropology is no more than a way of looking at the world and is notoriously content to appropriate other ways of doing so. Whatever the reasons, he threw himself into historical work. A piece he contributed to *All Souls under the Ancien Régime* (edited by S. J. D. Green and Peregrine Horden, 2007) thus did what a historian of Pisticci might have done. Never mind great books and intellectual currents, who *were* 'founder's kin' between 1600 and 1850? What were the kin links, what did these people own, who did they know and thus what *was* an Oxford college before the Victorian reforms? A reviewer referred to it as 'perhaps the most impressive and enlightening essay of the collection'.[49] John, though, adopted a self-aware pose of antiquarianism, and set about one of the College's parochial treasures, 'The Warden's Punishment Book'.

John retired in 2008. He is remembered as an efficient Warden, collegial but able to get things done. He is recalled as caring for the young, who can easily feel squashed by the 'big beasts', and he is remembered for his confident twitting of the old and the established when they needed twitting. At some point he had translated, rather beautifully, Carlo Cipolla's 'Fundamental laws of human stupidity' from *Allegro ma non troppo*. This, and Schopenhauer's 'Thirty-eight ways to win an argument', he sent as a 'welcome package' to those incoming heads of Oxford colleges whom he thought

[48] J. Davis, 'Administering creativity', *Anthropology Today*, 15/2 (1999), 7.
[49] M. Tworek, Review at H-education, H-net reviews in the humanities and social sciences, September 2010, http://www.h-net.org/reviews/showrev.php?id=30109 (accessed 1 May 2018).

showed signs of promise, though we have not ascertained when the practice started. Far less do we have a list of those who failed to receive a copy.

VIII

From his early days at LSE, John had moved steadily through a *cursus honorum*: the Curl Prize in 1969, the Malinowski Lecture in 1973, Honorary Secretary of the Association of Social Anthropologists and membership of the SSRC's committee on the subject through the early 1970s; in the 1980s membership of the Economic and Social Research Council's Research Resources and Methods Committee; chairmanship of the European Association of Social Anthropologists in the 1990s, and then President of the Royal Anthropological Institute. In short, all the things he felt Paul Stirling had failed to achieve but which Stirling may have thought irrelevant. Circumstances had changed, of course. John's career spanned a period when British anthropology had gone from a productively unsupervised state of opportunity, through a time of brutal cuts in funding, to a state of pettifogging regulation quite foreign to John's sensibilities. He himself emerged intact. From his days as a student, through his long career at Kent to his years in Oxford, he found himself in settings where a free exchange of thought and talk among well-informed friends was a practicable ideal.

A canting moralist—which John, decidedly, was not—would find the roots of later decay in earlier enjoyment. John had put on a great deal of weight, and by the time he retired he was not a healthy man. But he was as busy as ever, if not primarily with publication (*The Warden's Punishment Book*, co-edited with Scott Mandelbrote, came out in 2013), then with sociable encouragement of academics he thought well of. His ground-floor flat at Iffley, giving onto a garden, was lined with books and littered with papers, and his kitchen was as much a delight as ever. An evening of wine, *pasta all'amatriciana* and talk was a very civilised occasion. Anthropology had its place, but really that place was defined by literature, music and what used with more confidence to be called 'culture'; the greats of this or that academic subject might be acknowledged, although often deflated through some scurrilous story, but they were not to be venerated. It was rare on these occasions not to make a mental note to jot down afterwards some reference, anecdote or *bon mot*.

To what would have been the horror of some of his friends, had they known of it, John voted 'leave' in the Brexit referendum of 2016. He did so, as he explained the matter, from sheer dislike of the established powers telling him what to do. More than once in his Libya book he had mentioned with sympathy a man surnamed Bu Riziq who, with a small flock of animals, a pension and some grants from government (apart from the goats, in fact, very much like an ageing British academic), spent much of the year in a desert camp miles from anywhere. He knew what was going on in the world—he had his short-wave radio—but needed space of his own. Why did he live where he did? 'The last time a policeman came here was in 1936; very few people know what happened to him.'[50] Bu Riziq had no ambitions against the state. John's natural habitat was not a tent in the desert; a well-furnished house with a library, food and good wine was more his mark. But in a comfortable way he found space to do what he wanted.

John's influence was felt subtly through his interactions with colleagues. His anthropology was deliberately 'middle-range' and eclectic, and as a stylist he was succinct and understated. His writing could be crystal clear, but occasionally allusive, and he rejoiced in pedantic eccentricities such as beginning summative sentences with 'So …'. His speech, meanwhile, was littered with the mockingly Italianate 'My dears', which, addressed to male interlocutors in particular, was thoroughly distinctive. In some ways reserved and modest to a point—his *Who's Who* entries and CVs were brief, almost casual—he treated friendship, which mattered to him greatly, as a public performance. He could move from intimacy in cultivating persons he deemed import-ant to the social snub, and throughout his academic life was a master of the skilled put-down. But when relaxed, he was the consummate companion. Among friends he overcame his shyness and found the reassurance he needed to do his intellectual work.

Acknowledgements

The authors would like to thank Dymphna Hermans for comments, and also thank several colleagues who knew John well.

Note on the authors: Dr Paul Dresch is an Emeritus Research Fellow at St John's College, Oxford. Roy Ellen is Emeritus Professor of Anthropology and Human Ecology at the University of Kent; he was elected a Fellow of the British Academy in 2003.

---

[50] Davis, *Libyan Politics*, p. 154.

Doreen Barbara Massey

3 January 1944 – 11 March 2016

elected Fellow of the British Academy 2002

by

ROGER LEE

Biographical Memoirs of Fellows of the British Academy, XVII, 145–178
Posted 27 September 2018. © British Academy 2018.

DOREEN MASSEY

In his remarkable book *Blind Spot*, the photographer, writer and critic Teju Cole writes of A 'fluency in the dialect of geography'.[1] Although referring to the US poet Elizabeth Bishop whose work is characterised by a concern for the small everyday things of the world, especially scenes of work and production, to the Italian photographer and artist Luigi Ghirri who captures the oft-times surreal in landscape photographs examining the relationship between people and their environment, and to Italo Calvino the Italian journalist and author of stories of the wonders of the ordinary frequently heightened by the displacement of their protagonists, it is difficult to imagine a more appropriate—if terse—description of Doreen Massey's powerfully formative intellect, imagination and practice.

What is more, such connections with the visual and literary arts are especially apposite: Doreen's take on the world transcended particular disciplines and spheres of life to embrace them all. A clear manifestation of this inclusive understanding of the world was her founding in 1995 (along with Stuart Hall and Michael Rustin) of *Soundings*—a journal 'working within the Gramscian tradition of conjunctural analysis as a way of understanding the intertwining of complex forces in any given political moment'.[2] *Soundings* includes, for example, frequent close engagements with architecture, art and artists as well as a continuing fascination with the natural environment. And, above all, perhaps, it also insists on relationships between the quotidian and the extraordinary. However, *Soundings* was—and still is—primarily a left political journal. Its point, to show the wide relevance of left thought to contemporary life and to bring that thought to bear on the current political conjuncture.

Doreen's world was multi-dimensional and could, perhaps, begin to be understood only through the similarly multi-dimensional and open-ended qualities of Geography as a discipline. Her life was lived in and through Geography (as an academic discipline) and geography (as lived spatial relations). However, although this assertion is entirely appropriate, it is all too easily reductive and limiting. Teju Cole refers to what he calls 'the shimmering boundary between the map and the territory'[3] but for Doreen such a boundary shimmered with the formative relationships between them. Furthermore, the spatiality often associated with the notion of 'dialect' opens up the frame of the dialectic—with which she was thoroughly and critically aware—but, as befits her (geographical) world-view, it does so in a way which recognised and incorporated the kinds of multiples associated with dialects. The relationships between dialectic and dialect are an important reminder that Doreen's over-riding intellectual

[1] T. Cole, *Blind Spot* (London, 2017), p. 24.
[2] Along with Stuart Hall and Michael Rustin, Doreen was one of the three founding editors. Not only conjunctural in the Gramscian sense, *Soundings* also publishes politically activist material across the literary and visual arts: https://www.lwbooks.co.uk/soundings/about-soundings (accessed 20 June 2018).
[3] Cole, *Blind Spot*, p. 24.

and political motivation was always with the ways in which geography shapes society and, thereby, informs politics. Her multi-dimensional interests both shaped and were shaped by this politics and geography.

On this she was insistent. Her notion of geography was simultaneously territorial and relational. It recognised that territoriality and relationality were in continuous but forever changing and mutually formative relationship. Social and environmental life can be lived only through place-based spatial relationships. On the one hand, these relationships are distinctively, and always dynamically, formed as 'territory'—with all that term implies about the interrelationships between the natural world and the social. But at the same time, places are also continuously shaped and reshaped by, as well as being formative of, the multitude of active links (flows of people, ideas, money, commodities ...) which extend across and well beyond them. Geography and geographies of life shaped by it are, thereby, heavily and influentially involved in the ways in which social, economic and political practices are conducted and, literally, *take place*. This notion of Geography simply cannot allow simplistic formal notions of dialectical reasoning to prevail. Geography and geographies are just too complex for that. They are inherently open—hence their vital formative influence in shaping not only social and natural science beyond Geography but of politics. 'Geography'—in Doreen's phrase—'matters'.[4]

And for her this is no form of idealism. It reflects rather the geographical recognition that multiple alternative possibilities are always present, are continuously being formed and are, therefore, always of political interest and potential. But it also reflects the responsibility of the geography of social life—a responsibility to recognise the geographical constitution of injustice and to expose, understand and resist that which is unjust. The inherency of political responsibility within the making of geographies was a central trajectory of her work.[5] Politics then is always open. It is so because the geographies—territorial and relational—through which lives are lived and unevenly developed imply a responsibility for political relations—for the relations of power continuously produced and embedded in uneven geographies and for the possibility of their transformation. An example of the practical implications of this conception of a formative responsible geography and politics and of Doreen's influence on public policy was the deal between the Greater London Authority under Ken Livingstone and Venezuela whereby the former exchanged expertise for the

---

[4] D. Massey and J. Allen (eds.), *Geography Matters! A Reader* (Cambridge, 1984).
[5] See, for example, D. Massey, 'Geographies of responsibility', *Geografiska Annaler, Series B, Human Geography*, 86 (2004), 5–18; D. Massey, 'Politicising space and place', *The Scottish Geographical Magazine*, 112 (1996), 117–23.

latter's cheap oil—a deal later scrapped by Livingstone's right-wing successor as Mayor of London.[6]

'... the pleasure in all this complexity also then asks you to be responsible towards it'[7]

And it is here that the inseparability of Doreen's politics, geography, Geography and life is rooted. Her (geographically heightened) awareness of difference and of the 'strong sense of injustice'[8] associated with it did not and could not allow a separation between her Geography and her politics. As she put it herself:[9]

> The fact that I have the characteristics that I have is the result of the geographies within which I am set. And those geographies ... are all full of power.

Not for Doreen the mere application of Geography to politics nor, conversely, the political selection of things geographical for study on the artificial grounds that they are thereby rendered 'relevant'. For her 'everything was political'[10] and so the notion of 'relevance' would have no meaning as her politics was played out and, thereby, formed through geographies. Her work in Geography was driven by her politics and, at the same time, her political activism was always shaped by the formative influences of geography. Everything that she did formed, and was shaped by, this 'seamless world'.[11]

In related fashion, Doreen could not recognise any separation between 'work' and 'life'. The phrase 'work–life balance' would simply have been irrelevant to her. Her work was her life and vice versa. But at the same time, this characteristic is all too easily reductive when applied to Doreen. She was a totally collaborative—if very demanding—colleague, writer and teacher. It was difficult—verging on the impossible—to emulate her total commitment to her work and to the politics which it informed and from which it gained its dynamic. During a clearly very uncomfortable inquisition concerning the slow progress being made by one of her former PhD students she

---

[6] R. Meegan, 'Doreen Massey (1944–2016). A geographer who really mattered', *Regional Studies*, 51 (2017), 1285–96.

[7] D. Livingstone and D. Massey, 'Geography and geographical thought', in R. Lee et al. (eds.), *The Sage Handbook of Human Geography Volume 2* (London, 2014), p. 742; video at http://bcove.me/siljk2yb (accessed 20 June 2018).

[8] Interview with BL, 21 September 2017.

[9] *Secret City*, 2012: Secretcity-thefilm.com (accessed 20 June 2018).

[10] Interview with TM, 16 February 2017.

[11] Interview with RA, 2 May 2017.

commented, 'I do not expect anyone to work any harder than I do.'[12] Some comfort! A co-supervisor subsequently felt the need to visit the student at home to ensure that any consequent damage to morale was not too great. 'Going home' for Doreen 'did not mean doing something different.'[13] But equally, when chatting about a very well-known and influential academic, she regretted the tendency to use him as a role model as emulation would have been well-nigh impossible. 'When', she once asked, 'does he do the washing-up?'

At the same time, although she was 'never susceptible to the lure of bourgeois pleasures', was 'judgemental about not being serious' and even that having fun was always 'linked to being serious',[14] she was a vibrant, life-affirming and joyous companion. A former colleague remembered 'dancing with Doreen until dawn'[15] during an Open University residential summer school. Her capacity for uninhibited joy was realised most especially perhaps when, Manchester born and bred, she was able to watch her beloved Liverpool FC. Although this was a love born in part by childhood experience and her mother's approval of certain of the club's players, it was also informed by the uneven inter-urban economic and social geographies of the two cities. Her excursions to the Kop at Anfield were always shared with colleagues and friends not only because their company enhanced the pleasure but also because her long-standing and severe osteogenic problems and small physical stature made solo visits impossible. She 'wished that she could switch off'[16] and her visits to Anfield were amongst those moments which were as close as she got to be able to do so. And maybe she did not need to switch off. She was 'never off duty'[17] and hers was, after all, 'a life made with friends and her work'[18] comprised throughout by 'intense intellectual discussion and seriousness of purpose'.[19]

The vibrant empathy with a community of others—clearly demonstrated by her delight at standing on the Kop and joining in all the singing and ribaldry—was widely recognised and immensely valued by the myriad audiences with whom she engaged in places ranging from church hall and pubs on wet Tuesday evenings to monumental international conference venues and the highest corridors of power. 'Conversations' best describes the style of Doreen's innumerable talks and presentations which, even in the most formal of surroundings, were always open, direct, crystal clear and, no

[12] Interview with DC, 15 August 2017.
[13] Interview with RA, 2 May 2017.
[14] Interview with MH, 14 March 2017.
[15] Ibid.
[16] Interview with MB, 17 February 2017.
[17] Interview with MH, 14 March 2017.
[18] Interview with SC, 23 February 2017.
[19] Interview with MH, 14 March 2017.

matter how abstract the subject, consistently elucidated by accessible examples. Her deep understanding of what she was talking about and 'her way with words may have swept people along' with her arguments.[20] She would frequently ask questions of her audience or seek confirmation from someone that she recognised that the point that she was making was accurate and appropriate. She was, in short, totally engaged both with the urgency of the topic of her talk and her audience, however big or small:

> she was one of the most charismatic speakers I have ever heard. I remember her tiny frame absolutely filling one enormous lecture hall with energy and passion, extemporising from handwritten notes, intensifying the entire space. I can hear her voice now, and her laughter.[21]

Given all of this, it is hardly surprising that Doreen's work has not only had enormous formative political and geographical leverage throughout the world or that it has been so widely and incisively celebrated. The widespread institutional recognition of her contributions offers a reminder of the breadth and extent of her scholarly interests. Doreen was a founder Academician of the Academy of Learned Societies in the Social Sciences (1999) and a Fellow of the Royal Society of Arts (2000) as well as of the British Academy (2002). In 1998, she became the first woman to be awarded the Lauréat Prix International de Géographie Vautrin Lud—the so-called 'Nobel de Géographie'. She was awarded the Victoria Medal of the Royal Geographical Society (1994), the Anders Retzius Gold Medal of the Swedish Society of Anthropologists and Geographers (2003) and the Centenary Medal of the Royal Scottish Geographical Society (2003), and received honorary doctorates from the Universities of Edinburgh, Glasgow, the National University of Ireland and Queen Mary, University of London.

Two substantial books (a *Festschrift*[22] comprised of eighteen chapters—and, so reluctant was Doreen to be honoured in this way,[23] compiled largely in secret to mark her retirement in 2009 as Professor of Geography at the Open University, a post that she had held for twenty-seven years; and a collection of more than twenty-five essays reflecting on 'a critical life'[24]) are the most formal of these responses. In the light of all of this material, to say nothing of the revealing biographical interview with Doreen conducted

---

[20] Interview with DC, 15 August 2017.

[21] G. Rose, *Remembering Doreen* Blog 12 March 2016: https://visualmethodculture.wordpress.com/ 2016/03/12/ remembering-doreen/ (accessed 20 June 2018).

[22] D. Featherstone and J. Painter (eds.), *Spatial Politics: Essays for Doreen Massey* (Chichester, 2013).

[23] Interview with DC, 15 August 2017.

[24] M. Werner, J. Peck, R. Lave and B. Christophers (eds.), *Doreen Massey: Critical Dialogues* (Newcastle upon Tyne, 2018).

as part of her 1998 *Hettner Lecture* in Heidelberg[25] as well as three outstanding and lengthy surveys of her work[26] and an extended appreciation of her major philosophical work *For Space* published in *M@n@gement*, an international journal of strategic management,[27] there is little point in rehearsing—far less knowledgably, it would have had to be said—its content here. What follows, therefore, is an attempt to record my own personal reflections on, and understanding of, her life—her geography, politics, birdwatching, football and companionship—set within the scholarly and political frames in which she lived and worked.

And this is an appropriate stance[28] to take as in many ways it was the 'Twitter storm'[29] that followed her death in 2016 that perhaps best reflects the immediate and widespread influence that Doreen had on countless individuals and the multifarious ways in which her life and work connected so directly with theirs.[30] Without exception, these responses display not only admiration and respect but genuine affection for someone who, even if she was not known personally to the authors of the Tweets, seemed somehow to be so known. The reason for this deep and widespread connection with Doreen was not only that 'her ideas ... have transformed huge swathes of human geography and beyond' but that 'she could ... be incredibly warm—to everyone and anyone'.[31] This warmth translated into deep and lasting friendships even when Doreen's political trajectory diverged markedly from those of her friends with whom, in at least one case, she continued actively to make many excursions: walking, visits to National Trust sites, eating out, holidays and modernising her tiny flat.[32]

[25] T. Freytag and M. Hoyler, 'I feel as if I've been able to reinvent myself—a biographical interview with Doreen Massey', in D. Massey, *Power-Geometries and the Politics of Space-Time: Hettner-Lecture 1998* (Heidelberg, 1999), pp. 83–90.

[26] F. Callard, 'Doreen Massey', in P. Hubbard, R. Kitchin and G. Valentine (eds.), *Key Thinkers on Space and Place*, 2nd edn (London, 2011), pp. 299–306; N. Castree, 'A tribute to Doreen Massey (3 January 1944–11 March 2016)', *Progress in Human Geography*, 40 (2016), 585–92; Meegan, 'Doreen Massey (1944–2016)'.

[27] B. Sergot and A.-L. Saives, '*Unplugged*—Relating place to organization: a situated tribute to Doreen Massey', Cairn.info.htm, 2016; https://www.cairn.info/revue-management-2016-4-p-335.htm (accessed 20 June 2018).

[28] 'Appropriate', certainly, but also ironic in the light of Doreen's reluctance to engage with social media.

[29] H. Wainwright, '"How we will miss that chuckle": my friend, Doreen Massey', reprinted in Werner et al., *Doreen Massey: Critical Dialogues*; see also https://www.opendemocracy.net/uk/hilary-wainwright/how-we-will-miss-that-chuckle-my-friend-doreen-massey; http://www.redpepper.org.uk/tribute-doreen-massey/ (accessed 20 June 2018).

[30] And this is to say nothing of the range of obituaries which appeared in the national and local media from *The Guardian* (27 March 2016) and *The Daily Telegraph* (21 March 2016), to the *Manchester Evening News* (13 March 2016) and the *Ham and High* (29 March 2016).

[31] Rose, *Remembering Doreen*.

[32] Interview with KG, 24 April 2017.

'You need passion and you need rigour too'

In a conversation videoed in 2013,[33] Doreen discussed ideas of geography and
geographical thought with the geographer David Livingstone whose work has focused
especially on the historical geography of ideas. During this conversation, Doreen
would not allow the discussion to stay simply at the purely intellectual level. She constantly drew out the political inherent in all intellectual practice and railed against the
over-use of the word 'critical' in contemporary geography.[34]

> Nobody ever does anything in geography that isn't 'critical' … [and] that evacuates the
> term of all meaning. … We should really have a think in Geography about what a
> really serious critical Geography might be, and in what sense Geography can be an
> intervention that *does* make a difference. And it isn't just establishing your own
> credentials as progressive and something to put on your CV. We've got to get beyond
> that . … You need passion and you need rigour too.

This passage sums up her approach to her work as a completely committed practice
inseparable from wider concerns in life, oblivious to mere self-advancement and issues
of academic or political identity, wholly serious, 'challenging and determined'[35] and
demanding 'intellectual rigour',[36] all of which might combine to make a political and
hence social difference. This was how she lived and worked—with rigour and clear
political purpose—and, whilst she did not expect more from her students and her
colleagues, she did not expect less.

In her blog post *Remembering Doreen* Gillian Rose wrote 'she wasn't always an
easy person to work with. She could be very critical; she could insist on things being
done her way'[37]—'everything on her own terms'.[38] And she would make that expecta-
tion perfectly plain, even coming across with a 'slight prickliness'.[39] Her 'intense moral
rectitude', even 'puritanism', ensured a brutal 'honesty' and the 'unvarnished truth'.[40]
She would, for example, never write an endorsement of a book as she saw such practices not merely as mere advertising for publishers and almost always less than fully
honest[41] but also as an often gendered form of nepotism.[42]

---

[33] Livingstone and Massey, 'Geography and geographical thought'.
[34] Ibid.
[35] Interview with MB, 17 February 2017.
[36] Livingstone and Massey, 'Geography and geographical thought'.
[37] Rose, *Remembering Doreen*.
[38] Interview with MH, 14 March 2017.
[39] Ibid.
[40] Interview with DC, 15 August 2017.
[41] Ibid.
[42] Interview with BL, 21 September 2017.

Writing with Doreen was 'intimidating—it made you think very carefully about how you communicate, she was a stickler for grammar and punctuation'.[43] And fearing for the misplacement of apostrophes and other grammatical infelicities, supervisions with her could be 'terrifying'.[44]

> The reality is that we [at the OU] were all a little bit intellectually awed by her. She did not shrink from using her powerful intellect when she thought it necessary, though only with those who deserved it and could cope with it.[45]

This powerful aura was apparent in other circumstances too. During visits to meet the almost exclusively male executives at a number of large British corporations whilst undertaking research on industrial restructuring, Doreen's diminutive stature and informal dress could so easily be patronisingly dismissed—and was—until she began the interviews when it became immediately clear to assembled executives that 'we're not going to mess with her'.[46]

The insistent drive to 'get to the bottom of things, to work things out'[47]—phrases that were repeated in almost every conversation during the research for this memoir —were clearly acquired very early in her life. As a child on the Wythenshawe municipal housing estate south of Manchester on which she was brought up, she lived with her parents and her sister who recalls a meeting, called and chaired by Doreen—then around the age of seven—and held in her bedroom, to discuss the important issues of birds and birdwatching in their small back garden. At issue was the role played by the bird food placed in the garden by their mother for its avian visitors. As with so many apparently mundane matters in apparently ordinary life, for Doreen birds 'were a means to become curious and to explore—they were a focus of thought'[48] and they remained an object of delight and a means of relaxation, though never without serious intellectual purpose, throughout her life. Everything she observed had potential significance. A family walk along The Street—the Roman road heading south-east from Buxton [*Aquae Arnemetiae*] south-east towards modern Derby—prompted the querying observation that 'it's got a bend in it',[49] thereby challenging the common notion that all such roads were dead straight. Self-confidence in her ability came only in 'what she worked out'. This helps explain her insatiable desire to 'ferret things

---

[43] Interview with TM, 16 February 2017.
[44] Interview with BM, 4 May 2017.
[45] Email correspondence with MH, 8 March 2017.
[46] Interview with TM, 16 February 2017.
[47] Interview with MB, 17 February 2017.
[48] Ibid.
[49] Ibid.

out'.[50] Always based on immaculate preparation and 'taking nothing at face value',[51] this relentless dedication and thoroughness fed all of her academic work which was characterised by deep but critical knowledge and understanding.

A thorough minute of the bedroom meeting was taken by Doreen and recorded in a notebook. This was a precursor of the many meetings at the Open University (OU) preparing course texts and selecting readings to go along with them.

> It was pretty clear in these debates that, while all our views were taken into account, Doreen was the *prima inter pares*. Woe betide anyone who put forward ill thought out or weak ideas. She would very strongly defend her positions and ideas and criticise those she thought were weak, misguided or wrong. Doreen could be a tough taskmaster and I think all of us eventually had our abilities sharpened and developed.[52]

The consequence of this was that for at least one of her colleagues, 'the OU was … without question the most stimulating intellectual environment I have worked in',[53]

'The way we are, and the way places are, is a product of
our interrelations with everywhere else.
And those geographies, those relations, … are all full of power'[54]

Of all the themes with which Doreen Massey engaged in her intellectual/political practice, the relationship between the two fundamental dimensions of existence—time and space[55]—was what both drove her reformulation of geographical space and informed her politics based on what she called 'power-geometry'.[56] Although she suggested that it was her anger at the downgrading of space relative to time both within philosophy and, beyond Geography, in 'the rest of the social sciences' where space was treated as 'a kind of residual dimension' that provoked this work,[57] her annoyance was generated not by narrow disciplinary defensiveness but, as ever, by the political implications of the downgrading of space to 'a kind of flat, inert given'. She wished, therefore 'to bring space alive, to dynamise it and to make it relevant, to

[50] Ibid.

[51] Ibid.

[52] Email correspondence with MH, 14 March 2017.

[53] Ibid.

[54] D. Massey, Secretcity-thefilm.com/wp-content/uploads/2012/09/Secret-City Transcript.pdf (accessed 20 June 2018).

[55] D. Massey, *For Space* (London, 2005).

[56] D. Massey, 'Power-geometry and a progressive sense of place', in J. Bird, B. Curtis, T. Putnam, G. Robertson and L. Tickner (eds.), *Mapping the Futures: Local Cultures, Global Change* (London, 1993), pp. 60–70.

[57] D. Massey, 'Doreen Massey on Space *Social Science Bites*', podcast, Social Science Space.html, 2013: https://www.socialsciencespace.com/2013/02/podcastdoreen-massey-on-space/ (accessed 20 June 2018).

emphasise how important space is in the lives in which we live, and in the organisation of the societies in which we live'.[58]

And this process of revival had to include but go beyond a mere economism: geographical space is far too complex to allow such a reductive approach. It is also too complex to allow any notion of de-territorialisation and the consequent displacement of place as a powerful component not only of geographical analysis but also of social practice. Places—territories which may be defined at a whole range of scales from the hyper-local to the international—offer the complex formative context of such practice. The particularities of place shape the material, environmental, political, historical, locational, … contexts within which social practice *takes place* and so frame the boundaries of the possible. Places thereby contribute to differentiation and so de-legitimate any theoretical presumption of sameness. And places enable and constrain social practice—an example would be the possibilities of external economies of scale in financial practice enabled by clustering in places like the City of London. In these ways, place undermines any notion of inert space and so is central to the dynamisation of space.

These ideas began to be formulated in some of Doreen's early work at the Centre for Environmental Studies[59] but they started to take on the shape that would inform all her subsequent work in her extensive empirical research on industrial restructuring in the UK undertaken with Richard Meegan.[60] The notion that spatial factors were somehow independent of economic and social practice and could, therefore, be analysed as separate and autonomous causes of uneven development and the location of economic activity was blown apart by this work. It showed how corporate strategies responded to diverse forms of uneven geographical development by adopting a range of practices of profit-driven restructuring thereby causing place and space to be (re)constructed. This work and the subsequent two years spent as a competitively-awarded SSRC Senior Research Fellow at the London School of Economics (LSE) led in 1984 to the publication of her first single-authored book—the immensely influential *Spatial Divisions of Labour* appropriately subtitled *Social Structures and the Geography of Production*. The book had a profound transformative effect within economic geography and urban and regional studies setting both off on new trajectories by problematising the nature and significance of corporate geographies of production and their responsiveness to, and reshaping of, territorial and relational geographies.

---

[58] Ibid.

[59] From the outset, Doreen was inherently critical of the assumptions underpinning much of the work of CES. See, for example D. Massey, 'Towards a critique of industrial location theory', *Antipode*, 5 (1973), 33–9, reproduced from a piece initially published in the CES working papers.

[60] See, for example, D. B. Massey and R. A. Meegan, 'Industrial restructuring versus the cities', *Urban Studies*, 15 (1978), 273–88.

Revealing the use of space made by, and the transformative effects of, industrial capital it rewrote the way in which geographical space was thought and analysed. No longer could it be thought as being merely a passive stage on which economic, social and environmental practices were played out.[61] Rather it was an influential and formative component of those practices. Furthermore, *Spatial Divisions of Labour* formed a vital stage in the development of Doreen's thought towards the notion that 'geography matters' in highly complex and politically challenging ways.

Places change. They do so through the interaction of internal and external relations in ways which not only constantly redefine what is external and what is internal to place but reshape places themselves. In so doing, the very forces and relations of change are transformed and then, in a complex and continuous iterative process, a further round of change is set off. This notion of the interrelationships between territorial and relational space is both space- and place-making. Flows of money, for example, circulate in and across places and so change them. But in so doing such flows are themselves used, changed, interrupted, controlled, redirected by the circumstances of place which then change in turn and so further change the flows. An example of this kind of thinking in Doreen's writing and developed in *Spatial Divisions of Labour* is the notion of rounds of investment associated with the dynamics of production and corporate organisation. Geographies of investment reflect the profit-seeking forms of organisation and structure pursued by capital and so reflect past rounds of investment which shaped places and constrained and directed current rounds. Place—territorial space—is, therefore, in continuous iterative interaction with relational space; the two notions are distinguishable in thought but inseparable in practice. Politically, this is a very powerful notion. However, a critique of it is that in seeking to project place and space on to a wider politics the specificities and sense of place may be downplayed.[62]

But, for Doreen, the analytical drive of such thinking is directed towards the politics of geographical relations. And she shows that it has profound political significance.

> If time is the dimension in which things happen one after the other, it's the dimension of succession, then space is the dimension of things being, existing at the same time: of simultaneity. It's the dimension of multiplicity. ... it is space that presents us with the question of the social. And it presents us with the most fundamental of political of questions which is how are we going to live together.[63]

---

[61] D. Massey, *Spatial Divisions of Labour: Social Structures and the Geography of Production* (London, 1984; second edition, Basingstoke, 1995).
[62] Interview with BM, 4 May 2017.
[63] Massey, 'Doreen Massey on Space *Social Science Bites*'.

Before exploring the connections in her thought and practice between place, space and politics, it is worth pausing to note that in suggesting that how we are 'going to live together' is 'the most fundamental of political questions', she offers an insight into her own politics, grounded, left-orientated, shaped by place and highly committed—she was, for example, sceptical of one of her colleagues 'for not being fully engaged with the struggle'[64]—but always democratic.

Notions of simultaneity, multiplicity, (inescapable) togetherness and responsibility are both highly geographical and highly political. And these two qualities are closely interrelated. First, privileging time over space and thereby 'denying the simultaneity, the multiplicity of space ... and turning all those differences into a single historical trajectory'—turning space into time—converts 'contemporaneous difference ... into a single linear history'.[65] In the context of global uneven development this denies the notion of simultaneity, multiplicity and responsibility. All places are conceived as following the same trajectory from underdeveloped to developed with the implication that all should, therefore, necessarily conform to the norms of the developed. Aside from ignoring the possibility that alternatives are always present, such a formulation also denies the notion of power.

> Space concerns our relations with each other and in fact social space, I would say, is a product of our relations with each other, our connections with each other. So globalization, for instance, is a new geography constructed out of the relations we have with each other across the globe. And the most important thing that that raises if we are really thinking socially, is that all those relations are going to be filled with power. So what we have is a geography which in a sense is the geography of power. The distribution of those relations mirrors the power relations within ... society ....[66]

And, of course, this line of reasoning points to the geographical production and shaping of power itself. By virtue of their position in relational space, places are unevenly endowed with power. The geography of personal accumulation created by the multitude of local housing markets offers one obvious source of such power whilst its ability to reproduce itself within dynamic local markets serves to intensify its unequal distribution. Such unequal geographies of power operate at all scales from the local to the global. At the global level think, for example, of the power projected around the world by financial centres like the City of London[67] and spaces of offshore finance or by the headquarters of trans-national corporations and, at the national and

---

[64] Interview with MH, 14 March 2017.
[65] Massey, 'Doreen Massey on Space *Social Science Bites*'.
[66] Ibid.
[67] D. Massey, *World City* (Cambridge, 2007).

sub-national levels, of the power endowed upon place by the geography of national government and corporate geographies.[68]

Such spaces generate the real politics of social relationships and of 'living together' as well as shaping the formal politics of political allegiance and the struggle for power—a struggle with which Doreen Massey was engaged throughout her life. At the same time, her understanding of the underlying geographies of the formation and dynamics of politics and power also elucidates the kind of radical policies needed to redress inequalities of political power—policies that once again she sought to identify and put into practice.[69]

What is crucial in her 'thinking relationally about politics'[70] is not merely that the 'spatiality of society is part of what moulds and produces society—but also that we produce spaces that do have those effects'.[71] Not only, then, is it true that 'geography is integral to almost anything that happens'[72]—in other words that 'Geography matters'[73] in all the profoundly significant ways outlined above—but that those geographies and the power associated with them can be remade. And this implication of responsibility is an intrinsic and vital component of conceiving of geographical space in this way. Such a sense and practice of responsibility is both quotidian and endemic. It is politically inescapable.

'there was a consciousness that location mattered, that place mattered at a number of levels from the start'

In her conversation with David Livingstone,[74] Doreen described how she began to think geographically. Like Livingstone—a native of Northern Ireland—she was drawn to Geography and to the realisation of the inseparability of politics and geography by where she lived as a child.

---

[68] D. Massey, A. Amin and N. Thrift, *Decentering the Nation: a Radical Approach to Regional Inequality* (London, 2003). If *World City* addresses this issue at a global level, J. Allen, D. B. Massey and A. Cochrane, *Re-Thinking the Region* (London, 1998) does so at the regional/national level.

[69] Ibid.

[70] Interview with BL, 21 September 2017; see, for example, D. Massey, 'The conceptualisation of space and place and why it matters politically' (Norrkoping, Advanced Cultural Studies Institute of Sweden, 4 October 2007).

[71] Livingstone and Massey, 'Geography and geographical thought', p. 731.

[72] Ibid., p. 733.

[73] Massey and Allen, *Geography Matters*.

[74] Livingstone and Massey, 'Geography and geographical thought', p. 730.

> I come from Manchester, ... from the North (of England) and I was very conscious of
> regional inequality. It was part of our daily discourse. There was this saying that 'what
> Manchester thinks today London will think tomorrow'. Very much not just pride, but
> a kind of defiance about being dominated by another region.

The geographical and political significance of Manchester and its surrounding cotton
towns was manifest in a variety of ways not least, as Doreen and her colleague Linda
McDowell argued,[75] in their significance for the emergence of feminist politics
influenced by the centrally important role of women in the cotton trades:

> women in the north west played a significant part in the industrial history of the local-
> ity and so in the construction of that sense of place that distinguishes the north west
> cotton towns in general and Manchester in particular from other places.[76]

This is another profound and politically charged example of how and why
'geography matters'. And, for Doreen, it mattered too in a more local context:

> I also came from a council estate—corporation housing as we used to call it
> then—which had a really bad name. ... at times it was seen as a place that you wouldn't
> really want to put on your what we now call CV because that wouldn't help you to get
> a job.[77]

Notwithstanding this subsequent image and reputation of Wythenshawe, when
'Doreen was growing up in the forties and fifties, its reputation was perhaps best cap-
tured in the term "respectable"'.[78] And despite the facts that its design had been closely
influenced by the principles of the Garden City movement, espoused in Wythenshawe
especially by Shena Simon,[79] and that its rents were beyond the very low paid, Doreen
was always anxious to acknowledge the material benefits brought by the expanded
provision of state housing following the First World War. She came quickly to understand the crucial importance of what she called municipal socialism in the provision
of decent housing on open and well-designed estates of the sort exemplified by
Wythenshawe for a working-class population otherwise denied access to healthy

---

[75] L. McDowell and D. Massey, 'A woman's place', in D. Massey and J. Allen (eds.), *Geography Matters! A Reader* (Cambridge, 1984), pp. 128–47.

[76] L. McDowell, 'North and south: spatial divisions in a life lived geographically', in Werner et al. (eds.), *Doreen Massey: Critical Dialogues.*

[77] Livingstone and Massey, 'Geography and geographical thought', p. 730.

[78] McDowell, 'North and south'.

[79] Shena Simon was one of the three movers and shakers behind the development of Wythenshawe from the mid-1920s. Her husband, Ernest Simon, and the Labour Alderman W. T. Jackson were the others. Municipal Dreams in Housing, Manchester, *The Wythenshawe Estate, Manchester: 'the world of the future'*: https://municipaldreams.wordpress.com/2013/04/02/the-wythenshawe-estate-manchester-the-world-of-the-future/ (accessed 20 June 2018).

places in which to live.[80] After the Second World War housing densities on the estate were increased and the population grew to 100,000. Although Doreen felt alienated by the leafy fields south of the estate[81]—their incorporation into Manchester to provide the space for the estate had been resisted by rural Cheshire—no doubt their proximity helped to diversify the population of avian life studied so assiduously from her bedroom window.

And, from a very early age—and in an even more personally direct fashion—she came to value the immense redistributive power of the National Health Service created by the post-war Labour government. During her childhood she, her mother and sister made endlessly repeated visits to local NHS hospitals using municipal transport to enable the diagnosis and treatment of her life-long osteogenic problems. Told then that she would be in a wheelchair from the age of forty she successfully determined to resist that fate for the rest of her life.[82] This kind of resistant determination in the face of bodily fragility was also manifest in her drive 'to be the best'.[83] She also recognised the two-way process of involvement in the Welfare State in the form of responsibility and respect for it—a view that shaped her stance on the left of democratic politics.

If Manchester's historical geography was replete with powerful women, such women were a highly influential on Doreen's more immediate geography too. She grew up in a 'household of women'.[84] Her father was politically liberal—a gentle person who favoured discussion rather than confrontation. When not at work as the Groundsman of the Northern Lawn Tennis Club—founded in 1881 and moved south from Old Trafford to middle-class Didsbury at the turn of the century—for which he sought always to 'employ lads from local Labour Exchange',[85] he was a football referee and so often took himself off from the 'maelstrom of a house'[86] full of the laughter and debates of his wife and daughters. The family would often visit the 'the Northern' but 'always stayed only in the kitchen'—another marker of class position. In contrast to her father, Doreen's mother was an avid reader and wanted her daughters 'to improve' so teaching them manners and how to speak and to eat in order to enable them 'to get on'.[87] Much later, after they had both returned from University, she became a member of the Labour Party.

---

[80] Ibid.; McDowell, 'North and south'.
[81] T. Barnes, 'Her dark past', in Werner et al. (eds.), *Doreen Massey: Critical Dialogues*.
[82] Interview with MB, 17 February 2017.
[83] Interview with NG, 4 May 2017.
[84] Interview with MB, 17 February 2017.
[85] Ibid.
[86] Ibid.
[87] Ibid.

Just beyond the home—at the state primary school in nearby Sharston—the abundance of women teachers quickly recognised that, as they told Doreen's parents, 'you've got a very gifted child here'.[88] As such, 'perhaps she didn't realise that other people were not as clever'[89] and this difference may well have contributed to a degree of separation and independence in Doreen. However, her intellectual gifts continued to be fulfilled when she passed the entrance examination for the still highly prestigious (and now independent) Manchester High School for Girls—a school with many significant alumnae including Christabel Pankhurst—which Doreen was enabled to attend as a consequence in part of the boost to spending by the post-war Labour government on public education including an increase in the number of publicly funded places.[90]

Although suspicious of what she called 'origin stories',[91] Doreen's childhood and adolescent experiences offered a range of lessons in the significance of geography in the relationships of class and gender politics. It is not surprising then that there were 'no alternatives to Geography'[92] in her choice of subject for undergraduate study when she won a scholarship to St Hugh's College, Oxford in 1962. A major attraction was that the 'discipline itself opened up ways of thinking going well beyond the discipline'[93] and the tradition of the Geography taught at Oxford offered an integrative framework which informed not only her way of thinking but also fed the diverse interests that she maintained in the world ranging from geomorphology to feminist politics. Whilst the lack of theoretical depth may have made the subject less than intellectually stimulating, not least to Doreen, its broad and diverse content—unfamiliar to many geographers trained after the early 1960s—continued to inform her own understanding of a non-essential, non-foundational world. Geography, for her, was simultaneously both disciplinary and inter-disciplinary.[94] Indeed, the vital significance of the relationships between physical and human geography exercised her throughout her career.[95]

---

[88] Ibid.

[89] Ibid.

[90] McDowell, 'North and south', includes a number of evocative photographs of Doreen during her time at Manchester High School for Girls.

[91] Livingstone and Massey, 'Geography and geographical thought', p. 729

[92] Interview with MB, 17 February 2017.

[93] Interview with RA, 2 May 2017.

[94] Ibid.

[95] See, for example, D. Massey, 'Space-time, 'science' and the relationship between physical geography and human geography', *Transactions of the Institute of British Geographers*, n.s. 24 (1999), 261–76; S. Harrison, D. Massey, K. Richards, F. J. Magilligan, N. J. Thrift and B. Bender, 'Thinking across the divide: perspectives on the conversations between physical and human geography', *Area*, 36 (2004), 435–42.

However, the class and gender dislocations displayed and practised in Oxford were all too apparent: Doreen recalled, for example, being dubbed as 'only a mill girl'[96] on one occasion. This provocative confirmation of widely and deeply held class and gender positions and attitudes intensified, no doubt, by the journeys across deeply incised class boundaries as she moved back and forth between Oxford and Wythenshawe 'stung her into action'.[97] Arriving in Oxford as a 'working-class Tory',[98] she became, in her own words, 'a raving socialist',[99] her political radicalism stimulated by the realisation of the profound political significance of such attitudes and practices for systemic social injustice. And for the rest of her life, any boundaries between her politics and her Geography were simply non-existent.

Certainly, it is fortunate that Doreen's insatiable curiosity about the geography of the world around her prevented a debilitating disjunction between a highly traditional syllabus 'without passion and political commitment'[100] and the politics and social relations shaping the lives of young working-class women like her who, in the early 1960s, would have been even more exceptional presences in all UK universities let alone Oxford. Her comments on the Oxford experience[101] are more than a little reminiscent of those of Richard Flanagan twenty years later: Oxford was, he said, 'a finishing school for a certain class'.[102] However, despite her ambivalent response to her time at Oxford, in 2001 she became an Honorary Fellow of St Hugh's College and her undergraduate education did expose her to the breadth of geographical study and the complexity of the relationships that lie at its heart. Although she rejected an academic life à la Oxford,[103] her profound understanding of geographical complexity underpinned her intellectual approach to, and engagement in, social, environmental and political life.[104]

---

[96] Interview with MB, 17 February 2017.

[97] Ibid.

[98] Interview with BL, 21 September 2017.

[99] https://www.telegraph.co.uk/news/obituaries/12200242/Doreen-Massey-radical-geographer-obituary.html (accessed 20 June 2018).

[100] Barnes, 'Her dark past'.

[101] See, for example, Freytag and Hoyler, 'I feel as if I've been able to reinvent myself'; Barnes, 'Her dark past'; McDowell, 'North and south'.

[102] R. Flanagan, *Private Passions*, BBC Radio 3, Sunday 11 March 2018.

[103] Freytag and Hoyler, 'I feel as if I've been able to reinvent myself'.

[104] An excellent example of the geographically inscribed complexity of her understanding is offered in Allan Cochrane's discussion of the politics of London as a world city: A. Cochrane, 'Where is London?', in Werner et al. (eds.), *Doreen Massey: Critical Dialogues* (Newcastle upon Tyne, 2018).

'There is no way in which your gut politics or your gut ethical position should get in the way of an absolute clear, intellectual rigour …'[105]

Doreen graduated from Oxford in 1966 with a first-class degree and a Special College Prize and then—and in retrospect rather curiously—had a short-lived career in market relations before being employed as Principal Scientific Officer at the Centre for Environmental Studies (CES) by its then deputy director, Alan Wilson. CES had been set up in 1967 with a UK government grant as well as funds from the Ford Foundation as part of the then Labour government's drive to place scientific socialism at the heart of policy development.[106] Her early work at CES was heavily quantitative—and hence very different both from what preceded and what followed it. However, the abiding and motivating intellectual question in her mind throughout her life—'What does this mean?'[107]—and her single-minded concern to 'think things through' and to 'puzzle things out' enabled her to write a number of substantial and influential critiques of this work.[108] These qualities led to her appointment as an editorial assistant on the journal *Environment and Planning* (remaining as a member of the Editorial Board until 1986). And then, in 1971, they led her to take a year-long sabbatical at the Wharton School in the University of Pennsylvania to gain an MA in Regional Science with the doyen of the subject, Walter Isard, in order fully to understand the quantitative logic of her work at CES.

It is entirely without paradox that, whilst at Penn, she also took an elective course on French philosophy in French and Francophone Studies and so began a life-long interaction with the ideas of Louis Althusser who remained 'a key influence'.[109] The Althusserian notion of overdetermination remained with her, informing both her academic and political work. In this, it is possible to see the influence of the inherent complexity of geographical thought which, as she never failed to note, ranges across the physical, natural and sociocultural sciences and so is not reducible in any simplistic fashion. It helped inform her view of the world in which 'everything was important and fed into her bigger picture'.[110] Thus she rejected top-down structural frames of analysis and this led at times to forthright debates with others of a more essentialist cast. Of particular significance here is her response to the charge that studies of

---

[105] Livingstone and Massey, 'Geography and geographical thought', p. 741.
[106] Barnes, 'Her dark past'.
[107] Interview with MB, 17 February 2017.
[108] See, for example, Massey, 'Towards a critique of industrial location theory'.
[109] Interview with TM, 16 February 2017; Interview with DC, 15 August 2017.
[110] Interview with TM, 16 February 2017.

localities were unprogressive, even merely empirical. This Marxist critique[111] of what was a major multi-site programme in the UK, spawned under the influence of her work on spatial divisions of labour and funded by the ESRC during the late 1980s, elicited a sharp response from Doreen. Although she had not participated in the localities research programme and had reservations about what it delivered, the application of her thinking about territorial and relational space to rebut such arguments was devastating in its effectiveness.[112]

The linkage between the quotidian and the extra-ordinary was an inevitable and inescapable feature of her work. She 'would always get on the bus with the locals'[113] and so she comprehended the world as much from such apparently ordinary experiences as from her knowledge and love of philosophy: 'Much of life for many people, even in the heart of the First World, still consists of waiting in a bus-shelter with your shopping for a bus that never comes.'[114] As Emma Jackson puts it,[115] Doreen intrinsically recognised 'the need to keep one eye on global capital and another on the bus'. This distinctive quality of her thought—she was, perhaps, 'Marxian not a Marxist',[116] recognising with the likes of Raymond Williams, the ordinariness and profound significance of culture—is central to an understanding of her work and lies at the heart of the indivisibility of her academic and political practice.

This way of comprehending the world combined with her notions of the mutually formative links between relational and territorial space led to another of her insights—a global sense of place[117]—which transformed both the study of places and regions and of the causes and responsibility for global uneven development. All places are at once the product of their internal, yet dynamic, qualities in interaction with the host of influences from elsewhere that criss-cross and so change them, and of their projection of influence on and the constitution of other places. Whilst momentarily distinctive, places can never be self-contained. It is not possible to understand them simply in terms of their internal qualities no matter how distinctive these may be. And so it is wholly inadequate to try to understand regional uneven development, for

[111] See, for example, N. Smith, 'Dangers of the empirical turn: some comments on the CURS initiative', *Antipode*, 19 (1987), 59–66.

[112] For one example see her response to the Marxist critique of locality studies as unprogressive: D. Massey, 'The political place of locality studies', *Environment and Planning A*, 23 (1991), 267–81.

[113] Interview with MH, 14 March 2017.

[114] D. Massey, 'A place called home?', *New Formations*, 17 (1992), 3–15.

[115] Keeping one eye on the bus: a tribute to Doreen Massey, Emma Jackson.html: https://cucrblog.wordpress.com/2016/03/14/keeping-one-eye-on-the-bus-a-tribute-to-doreen-massey-by-emma-jackson/ (accessed 20 June 2018).

[116] Interview with TM, 16 February 2017.

[117] D. Massey, 'A global sense of place', *Marxism Today*, 35 (1991), 24–9.

example, in terms of regions themselves as they are always the product of relational as well as territorial processes.

These ideas were fully worked through and applied in her study of south-east England, *Rethinking the Region*,[118] jointly authored with colleagues from the OU. And this kind of analysis which implies an active form of responsibility was also applied to her understanding of uneven development at the global scale. Doreen was nothing if not an internationalist, recognising the power of globalisation not as a top-down set of processes but as engendered within and from places as relations of power—'power geometries' as she put it—which could, thereby, be resisted and countered. But she also recognised the potential for cultural imperialism in such work:

> A lot of us do geographies of other places. What we forget is there are geographers in those places and they do geographies of us, and there has been quite a lot of a kind of an imperial assumption of the dominance of our knowledge and cultural imperialism, knowledge imperialism.[119]

Not surprisingly, although she was pleased by the use of her notions of 'power geometry' in Hugo Chavez's Venezuela, she was fully aware of the need to be 'a lot more aware of the geographies being produced elsewhere'.[120] She was capable of writing in Portuguese and was fluent in French and Spanish—and was frequently to be observed brushing up on both languages—but, at the same time, she realised that her internationalism was inherently one-sided, limited by linguistic barriers. Thus, although she engaged closely with the Geography and politics of Venezuela, Nicaragua, Brazil and South Africa and made frequent visits to them, often being interviewed by the media, serving on the editorial boards of local journals and giving papers in the local language, her published output on these places was perforce more limited.[121]

Whilst her quantitative work at CES might appear to operate in a different world from her subsequent culturally inflected work on class and gender[122]—a cultural inflection which generated vigorous discussion with other more foundational critics[123] —and her continued close involvement in a variety of political causes, it was her interest in scientific socialism which appealed to her practice of thought and offered an integrative link between them. 'Thinking in a rigorous way' and 'analytic precision'[124] were inherent qualities of her work and she expected the same of others. In this

---

[118] J. Allen, D. Massey and A. Cochrane, *Rethinking the Region* (London and New York, 1988).

[119] Livingstone and Massey, 'Geography and geographical thought', p. 738.

[120] Ibid.

[121] See, for example, D. Massey, 'Nicaragua: some reflections on socio-spatial issues in a society in transition', *Antipode*, 18 (1986), 322–31.

[122] See, for example, D. Massey, *Space, Place and Gender* (Cambridge, 1994).

[123] D. Massey, 'Flexible sexism', *Environment and Planning D: Society and Space*, 9 (1991), 31–57.

[124] Interview with BL, 21 September 2017.

she pursued an intellectual trajectory similar in some respects to others—Bill Bunge, David Harvey and Kevin Cox are prominent examples[125]—who had moved from the inaccurate and wholly misleading precision of the so-called 'quantitative revolution' in Geography to a Geography informed by Marxist thought. However, although she admired the hyper-rationalist work of a number of sociologists, for example, this attraction to rationality was combined with 'a cultural sensibility'[126] no doubt derived in part from her non-essentialised geographical view of the world. Influenced by Althusser's notion of overdetermination, her receptivity to Marxian ideas reflected her own non-reductive thought. And this, in turn, was shaped by her geographically informed understanding of the complexity and historically and geographically variable nature of relationships within and between environmental and social life.[127]

Unsurprisingly, then, she was very keen to promote non-quantitative work within CES and she was instrumental in persuading the Centre to appoint Richard Meegan as a non-quantitative research associate with whom she undertook large-scale empirical work on a variety of highly influential and transformative research projects concerned primarily with the relations between industrial restructuring and the changing uneven geography of the British economy as well as with the methodologies appropriate for such investigations.[128] The publications from this research are classics of their kind whilst the book-length treatment of the relationships between politics and methodology—itself a typically Masseyan conjuncture—remains staple fare over thirty years after its publication.

Much of the writing-up of this empirical research supported by Doreen's SSRC-funded Fellowship was completed in an office close to Trafalgar Square after the closure of CES in Chandos Place. As with all her collaborative projects, the writing was shared, with Doreen's contributions all written out in long hand. 'Everything was hand-written' as this technology made for a 'more natural transmission of thought from head to arm to pen'[129] and it was then sent to Manchester to be typed by her aunt.[130] This remained Doreen's way of writing—with her aunt's role taken over eventually by secretarial colleagues.

---

[125] I am grateful to Felix Driver for reminding about these similarities.

[126] Interview with DC, 15 August 2017.

[127] Interview with RA, 2 May 2017.

[128] D. Massey and R. Meegan, *The Geography of Industrial Reorganisation: the Spatial Effects of the Restructuring of the Electrical Engineering Sector under the Industrial Reorganization Corporation* (Oxford, 1979); D. Massey and R. Meegan, *The Anatomy of Job Loss: the How, Why and Where of Employment Decline,* (London, 1982); D. Massey and R. Meegan, *Politics and Method: Contrasting Studies in Industrial Geography,* (London, 1985). This joint work fed directly into one of Doreen Massey's most influential papers published in 1979: 'In what sense a regional problem?', *Regional Studies,* 13 (1979), 233–43.

[129] Interview with RA, 2 May 2017.

[130] Ibid.

Her appointment to the Chair of Geography at the Open University in 1982 was not only a 'brave'[131] decision but innovative,[132] far-sighted and self-evidently highly effective. By that time, she had spent twelve years at the CES and two years as Senior Research Fellow in the Department of Geography at the London School of Economics as well as holding many part-time appointments, including several as an external examiner, a number of lecturing posts and acting as an external assessor at the Open University. Doreen had also produced many influential publications: five books including two edited collections, seventeen journal papers and nine chapters in edited books.[133] So maybe not so 'brave' an appointment after all? However, she had never held a full-time academic post, had no PhD and had only part-time teaching experience.

'The appointment undoubtedly appeared unusual to outsiders'. However, despite caution and even timidity amongst the higher echelons of the OU, 'within the faculty it fitted well with the exciting sense that we were building an innovative group of collegial researchers and teachers with the desire and ability to break old boundaries between and within disciplines. Indeed, that had been one of the founding tenets of the OU as a whole and had been successful in other faculties.'[134] Stuart Hall had joined the OU in 1979 and Laurence Harris in 1980 whilst David Potter was centrally involved in 'creating a ground-breaking interdisciplinary foundation course'[135] in Social Science. When it became apparent that Doreen might be attracted to the post, it was not only the 'power brokers'[136] at the OU who were enthusiastic but also the group of young-ish social geographers on the faculty keen to develop the relationships between Geography and other social sciences. Nevertheless, the university 'made the appointment conditional upon Doreen passing a medical'.[137]

Loving 'team work and being around ideas', she was 'in her element at the OU' whilst at the same time the structure and processes of learning and teaching at the OU enabled her 'to prepare very well for everything as she always needed time to think'.[138]

---

[131] Interview with RA, 2 May 2017.

[132] Doreen was only the seventh woman ever to be appointed to a Chair in Geography in the UK: see R. J. Johnston and E. V. Brack, 'Appointment and promotion in the academic labour market: a preliminary survey of British University Departments of Geography, 1933-1982', *Transactions of the Institute of British Geographers*, n.s. 8, 100–11. I am grateful to Felix Driver for drawing my attention to this. It reveals just how iniquitous was the gendered academic appointments process as well as demonstrating its stupidity and capacity for self-harm.

[133] For a selection of her publications see B. Christophers, R. Lave, J. Peck and M. Werner (eds.), *The Doreen Massey Reader* (Newcastle upon Tyne, 2018).

[134] Email correspondence with SO, 16 March 2018.

[135] Ibid.

[136] Ibid.

[137] Ibid.

[138] Interview with NG, 4 May 2017.

Before writing anything, everything—structure, direction, argument, coherence—had to be thought through. Nothing would be committed to paper until she had the whole fully secured in her mind. For these reasons she would have found the day-to-day uncertainties of face-to-face teaching—to say nothing of the vicissitudes of administration and management—very difficult to deal with: 'she was not a good manager as she could become obsessed about stuff which did not matter'. But 'she was a leader': she 'shaped the agenda and way of thinking'.[139]

However, although 'it was very clear who was the intellectual leader'[140] and that 'she was often way ahead of the rest of us',[141] she always accepted her position as one member within the course team. The parameters of an essentially collaborative project and group discussions 'always involved negotiation'.[142] She wanted to ensure that the courses dealt with issues that were socially and politically important. But in her identification of importance, issues of ethnicity and religion were more awkward to deal with then those of class and gender—although her analysis of class was made from a feminist perspective.[143] So there was always 'a lot of intellectual struggle about whether certain issues should be dealt with'. 'One could never wholly relax' and, despite her ready sense of humour, one always 'had to be alert'.[144] She could be 'abrasive', even 'combative', and 'she was very demanding especially of people close to her but who maybe led more complicated lives'.[145]

Although working with others was central to her modus operandi—and the legendary informal hour-or-so-long seminars journeying between north London and Milton Keynes by car or train with the likes of Stuart Hall and Laurence Harris were occasions of intellectual wonder for postgraduates making the same journey in the same vehicle[146]—she also worked long and hard on her own in her 'little flat',[147] the windows of which she famously regarded as her most treasured possession.[148] Here was another space—the flat itself located in a district, Ariel Road, in which she felt at home and at ease as she 'loved the mixture of Kilburn'.[149] Echoing her relationship with Wythenshawe, Kilburn formed a community around her from which she gained

---

[139] Interview with RA, 2 May 2017.
[140] Interview with MH, 14 March 2017.
[141] Interview with RA, 2 May 2017.
[142] Ibid.
[143] Interview with DC, 15 August 2017.
[144] Interview with MH, 14 March 2017.
[145] Interview with RA, 2 May 2017.
[146] Interview with DC, 15 August 2017.
[147] Interview with SC, 23 February 2017.
[148] R. Lee, 'Doreen Massey', *The Geographical Journal*, 182 (2016), 311–12.
[149] Interview with SC, 23 February 2017.

crucial quotidian 'feedback'.[150] The flat had been acquired via a mortgage scheme backed by the London Borough of Camden[151] and, as with the complete lack of separation for Doreen between life and work, it was a place of work as much as of residence. It was a frugal—even 'ascetic'[152]—space leavened with some art[153] but lacking many everyday items such as, for example, a TV which she was eventually cajoled to acquire[154] but only after many years of abstinence—of which, of course, she was unaware.

It was from this space that all her own writing emanated and in which much of the politics—and especially the re-framing of debate on the left—in which she was engaged was conceived and developed. Although academic and political colleagues would visit Ariel Road for work—to write joint papers and books, establishing *Soundings* and the *Kilburn Manifesto*, for example—much of her work involved applying her 'fierce concentration'[155] to whatever she was currently thinking, reading or writing. Her insistence on working things out thoroughly for herself led to a profound sense of academic integrity—'she was never a jumper on a band-wagon'.[156]

The diverse positive effects of this intensity and demanding seriousness of purpose and independent style of collegiality were manifest in other ways too. The remarkable range of course-based publications emerging from the collaborative work at the OU were distinguished by a singular style—highly innovative in conceptual and theoretical terms, yet always grounded in material realities and crystal clear in expression. Whilst the work involved in these publications served to reveal the ridiculous contradictions embedded in Research Assessments in that it may well have limited the OU Geography's research rankings,[157] they also had remarkable transformative effects on others both near—'I could never have made the move from planner to academic without the work with Doreen',[158] 'she got us to do stuff that we didn't want to do and think things we didn't want to think'[159]—and far in their use by the students of the OU and in their adoption by many other universities around the world.

The dedication to, and dedication of, her work was far from being confined to the OU. In her close involvement with *Environment and Planning*, *New Left Review* and then *Soundings* and the *Kilburn Manifesto*, Doreen was 'incredibly hard-working' and

---

[150] Interview with MB, 17 February 2017.
[151] Ibid.
[152] Interview with KG, 24 April 2017.
[153] Ibid.
[154] Interview with TM, 16 February 2017.
[155] Interview with BL, 21 September 20i7.
[156] Interview with DC, 15 August 2017.
[157] Interview with SC, 23 February 2017.
[158] Interview with TM, 16 February 2017.
[159] Interview with RA, 2 May 2017.

'scrupulous'.[160] And this degree of commitment was necessary as the *New Left Review* was 'a bit of a fiasco'[161] in the late 1980s and early 1990s when Doreen became a member of the editorial committee whilst *Soundings* and the *Manifesto* were co-created by her from scratch.

Being Doreen Massey: 'a public figure in a private space'[162]

Two central characteristics of 'being Doreen Massey'[163] were her continuing wonder at the world around her and her sense of responsibility for it. Both were strongly shaped by the powerful intersections of geography and politics through which she understood the world. This wonder and responsibility were always closely placed in her work. Not for her a displaced view of the world from some distanced or detached prospect. She always saw the world from somewhere—a somewhere which was itself always fully located in territorial and relational space. She was 'committed to locality'.[164] This meant that her politics and geography were always thoroughly grounded and so subject to the transmission of meaning. They were thereby directly related to specific material realities rather than some generalised version of such. For her, a place-based class politics is 'generative'.[165]

If Wythenshawe and Ariel Road, Kilburn, were two especially meaningful places for Doreen, so too was Anfield—the home of Liverpool FC. 'No moments were wasted'[166] in her life and so football—the songs, the shouting, the spectacle, the community of fans on the Kop—offered a joyful release. At the same time, of course, for Doreen the experience was soaked in class politics and geographical identity. Communal participation in football was a 'common currency'[167] combining politics and geography. For Doreen who 'could talk to anybody',[168] the football terraces were another place that enabled her to connect directly with the world around her.

Although she did not wish to be a celebrity, she was keenly aware of the extent to which her work had been formatively influential and filed her papers in the form of an

[160] Interview with SC, 23 February 2018; Interview with BL, 21 September 2017.
[161] Interview with SC, 23 February 2018.
[162] Interview with NG, 4 May 2017.
[163] Interview with MB, 17 February 2017.
[164] Interview with SC, 23 February 2017.
[165] Interview with BL, 21 September 2017.
[166] Interview with SC, 23 February 2017.
[167] Ibid.
[168] Interview with KG, 24 April 2017.

archive for posterity.[169] She 'fed off'[170] the reactions to and enthusiasm for her public appearances and she coped with the demands of fame and her active involvement in public debate with the sense of responsibility which informed her politics—trying always, for example, to reply to all the contacts made with her and realising only gradually 'that she could say "No"'[171] to at least some of the countless invitations that flowed constantly in her direction. She was a consistently active and constructive spokesperson for the discipline of Geography and, notwithstanding the immense reach and influence of all the OU courses on which she worked, far from only at the level of higher education and research.[172] However, she did consider her retirement very carefully and was clear that whilst the intellectual engagement with politics would most definitely continue, 'she had no thoughts about carrying on with academic work'.[173]

Although she was greatly influenced by the writings of Louis Althusser and by other colleagues, not least Stuart Hall, Chantal Mouffe and Ernesto Laclau, as well as occasional group discussions with academics when their interests coalesced with hers,[174] and although she sustained close political relationships with the likes of Ken Livingstone and Ed Miliband she always maintained a degree of separateness.[175] The so-called Ariel Road Group was a loose grouping including amongst others Stuart Hall, Michael Rustin, Hilary Wainwright, Robin Murray and Ken Livingstone, the discussions of which—much informed by the political consequences of the abolition of the GLC—led to the formation of *Soundings* and, later, to the *Kilburn Manifesto*. But although these groups were self-evidently powerfully formative and vital to her, she did not ally herself definitively with a particular group of activists or scholars. So she retained a degree of independence—one of her close collaborators described her as 'quirky and independent'[176]—and was able to work productively with others on specific issues on which they found common ground whilst at the same time she might disagree with them fundamentally on wider issues and political positions.

In short, 'whilst being an intellectual' her approach to political activism was shaped by a desire to retain 'a chance to have a direct input to strategy' by identifying

---

[169] Interview with NG, 4 May 2017.

[170] Interview with SC, 23 February 2017.

[171] Interview with KG, 24 April 2017.

[172] Her active involvement in the support and progressive transformation of Geography in schools was recognized when she became Honorary Vice President of the Geographical Association (1989–1993).

[173] Interview with RA, 2 May 2017.

[174] Ibid. Such groups were loose and often labeled by using the names of the places from which their members came or the routes they had to travel to get to the places in which they met.

[175] See, for example, D. Massey, 'The world we're in: an interview with Ken Livingstone', *Soundings: a Journal of Politics and Culture*, 36 (2007), 11–25.

[176] Interview with KG, 24 April 2017.

a 'feasible left alternative'.[177] This stance informed her collaboration with Ken Livingstone in the Greater London Enterprise Board and the development of the *London Industrial Strategy*, for example, and with the Labour Party in which she served on a number of sub-committees of the National Executive Committee as well as with her drive to produce the *Kilburn Manifesto* and to establish *Soundings*. They reflect Doreen's role as a public intellectual—not so much an activist—and her desire to rethink and energise left politics through intense intellectual and practical engagement with questions not only of political thought—as in *Soundings*, for example—but in the *Manifesto*, with the detailed development of practical, left-orientated and post-neo-liberal policies for all the spheres of activity of a modern state.[178] Both of these publications are ongoing.

However she also 'needed political heroes' and was, perhaps, 'unduly kind and blind to the weaknesses' of those who, like Hugo Chavez, enthusiastically understood the world politically in line with her own views.[179] She was 'always conscious—and self-aware of needing to be involved in the struggle of the oppressed'.[180] At the same time, however, and, although she was an 'uncompromising and staunch socialist' and 'class politics' formed the bedrock of her political position—a position that put her at odds with some of her academic collaborators[181]—she was fully aware of the difficulties of pursuing socialism within the context of a democracy in which left-wing views are so easily demonised and 'would not have been for political suicide'[182] by rigidly adopting positions which seemed to undermine the possibility of electoral legitimacy.

And this put her at odds with some of her political collaborators too.[183] Rather than wishing to be directly politically active, she wanted to 'be in a position to make a difference' to influence the political community to which she belonged by 'thinking politics differently'[184] through geography, 'searching for the most appropriate form of organisation for the struggle' and by 'trying to work out the most appropriate way to draw together different social groups'[185] rather than to impose a particular way of thinking and acting. The *Kilburn Manifesto* was a manifestation of this urge to 'articulate a different way of thinking for the left'.[186]

---

[177] Interview with MH, 14 March 2017.
[178] D. Massey, S. Hall and M. Rustin (eds.), *After Neoliberalism? The Kilburn Manifesto* (London, 2015).
[179] Ibid; Interview with RA, 2 May 2017.
[180] Interview with MH, 14 March 2017.
[181] Interview with DC, 15 August 2017.
[182] Interview with SC, 23 February 2017.
[183] Interview with RA, 2 May 2017; Interview with DC, 15 August 2017.
[184] Interview with BL, 21 September 2017.
[185] Interview with MH, 14 March 2017.
[186] Interview with BL, 21 September 2017.

And this stance was equally true of her academic work. Characterised less by empirical enquiry than by (re)interpretation, she would read such research 'and tell you what it said' in her own terms. 'She thought in different ways about stuff you'd taken for granted' and she thereby 'shaped research agendas and ways of thinking.'[187]

Because she had always worked things through so thoroughly and was, thereby, 'confident in what she worked out'[188] and was 'secure in her own sense of conviction'[189] she 'could cope with everything on a platform'[190] but she was less confident in other circumstances or even with friends 'on the other side of the tracks'—in 'the Hampstead or Oxford scene',[191] for example—and she found class difference and the 'social graces and self-confidence of the metropolitan middle class'[192] difficult to deal with, certainly at a personal level. She had 'zero interest in food' as a cultural signifier—seeing it as a 'bourgeois concept'[193] and 'didn't do cooking'[194]—at least in spending time thinking about food and cooking as a marker of consumption. However, she was more than well-aware that cooking good and simple food mattered and cut across relations of class. Attracted by its Partnership scheme, she shopped at Waitrose and was discriminating in her choice of wine when eating out.[195] However—and in large part because of her deep sense of personal responsibility to others and ethical use of her funds, not least in using her savings to fund the foundation of *Soundings*—she was 'a big saver and used a financial adviser'.[196] In a broader context—and although 'she did not suffer fools gladly'[197]—she was reticent in her perforce extensive dealings with the medical profession and was 'badly let down'[198] by the NHS in her treatment after one of the falls that incapacitated her for several months. But the deeper cause of such diffidence was that she did not put herself first. Facing several winter weeks with no central heating in the flat she put up with the delays to the necessary repairs saying that 'it was not that cold'.[199]

Although an essentially 'private person' and 'difficult to get to know'[200] her

[187] Interview with RA, 2 February 2017.
[188] Interview with MB, 17 February 2017.
[189] Interview with MH, 14 March 2017.
[190] Interview with SC, 23 February 2017.
[191] Ibid.
[192] Interview with MH, 14 March 2017.
[193] Interview with KG, 24 April 2017.
[194] Interview with SC, 23 February 2017.
[195] Interview with KG, 24 April 2017.
[196] Interview with SC, 23 February 2017.
[197] Interview with KG, 24 April 2017.
[198] Ibid.
[199] Ibid.
[200] Interview with BM, 4 May 2017.

friendships, most of which were rooted in political or academic relationships, were deep and loyal.[201] And they would remain so even in the face of a divergence of professional, academic and/or political interests. Her loyalty would remain despite her disapproval—which she would make perfectly plain—of, for example, the career choices made by certain of her colleagues.

Nevertheless, although she was 'very focused on geography and politics' she 'had a much wider hinterland' and so was not 'politically categorical in friendships'.[202] She loved travel and was always prepared to be able to indulge this passion often delighting in seeing at first-hand geomorphological features at which she had wondered in school and university textbooks. Journeys with Doreen were an education as she was able to maintain an almost continuous commentary on the physical and historical geography and the avian identity of the places encountered as the train or the car sped by the landscape.[203]

Although her flat was 'immaculate'[204] and a 'sanctuary'[205] for Doreen, it was only sparsely provided with home comforts and she had very few material possessions—most such did not enter her perception of relevance—but she was interested in design and the visual arts. At one point she was persuaded that it really would be a good idea to bring her tiny galley kitchen at least minimally up to date. And once committed to such a course of action Doreen applied the same ferocious commitment and concentration to it as she did to her wider work searching enthusiastically for appropriate wall tiles to bring some colour to it and as a reminder of times spent in Mexico. But the politics of domestic refurbishment was always present: Doreen held on to her old cooker and would not countenance the installation of expensive designer mixer taps.[206]

*****

And here we are back where this memoir began: with the small things of the world, with everyday scenes, especially of work and production, and with the wonders of the ordinary from which Doreen Massey developed a geography and a politics which transformed the ways in which space and time are thought. Far from being a much-travelled academic star—and despite the global significance of her work and its transformative effects on Geography, social science and politics—Doreen chose to hold

[201] Interview with KG, 24 April 2017.
[202] Ibid.
[203] Interview with KG, 24 April 2017; Interview with TM, 16 February 2017.
[204] Interview with DC, 15 August 2017.
[205] Interview with KG, 24, April 2017.
[206] Ibid.

only two long-term permanent posts throughout her career: the twelve years at the CES (1968–1980) and the thirty-four years as Professor of Geography in the Faculty of Social Sciences at the OU (1982–2016), the last seven years as professor emeritus. These two posts were separated by a two-year stint as an SSRC Senior Research Fellow at the LSE. Once again, the importance for her of working from place is apparent in this relatively stable geography. The span from the quotidian to the extraordinary—or rather, perhaps, from recognising the extraordinary in the quotidian—is what enabled her remarkable Geography and her politics.

Her always grounded understanding and ways of thinking and communicating transformed theory and practice in contemporary geography. Beyond that she insisted on the spatialisation of social science, thereby transforming thinking about the nature of place, space and time. Her internationalism was inherent in the way in which she thought about the world, but she was never imperialist and was insistent that making geographies always entailed direct, immediate, personal and political responsibility. Through her insistence that geography and politics are not only inseparable but mutually formative, she changed the way in which politics is thought. Widely honoured by major scholarly societies and institutions, she refused the award of an OBE.

And all of this took place within the context of her graduating in an essentially descriptive Geography founded in the idiographic. Despite several decades of quantitative locational research emanating from scholars such as August Lösch in Germany and taken up by regional scientists—including Walter Isard—and geographers in the USA and Sweden, the idiographic approach in Geography remained predominant. By the mid-1960s, however, it was under serious challenge. Its predominance was about to be radically—even violently—transformed by what became known as spatial or locational analysis. Peter Haggett's book *Locational Analysis in Human Geography* was published in the UK in 1965, the year of Doreen's graduation. This book mounted a fundamental challenge to idiographic Geography—in some ways replacing Geography with Geometry. Whilst elements of locational analysis had by the mid-1960s begun to find their way into urban geography (central place theory) and economic geography (industrial location theory and network analysis) it had by no means revolutionised or displaced the thinking underlying idiographic Geography which remained concerned with the nature of the geographical—essentially regional—unique and so 'with asking biographical questions about the phenomena we observe'.[207] This changed dramatically from the mid-1960s onwards with the geographical pursuit of what Haggett called 'the search for order … asking questions

[207] Ibid.

about the order, locational order, shown by the phenomena studied traditionally as *human geography*'.[208]

It was with the ideas and objects of analysis deriving from locational analysis that Doreen worked at the outset of her career in CES. And, as her sojourn in Pennsylvania testifies, she took this approach seriously—even if only the better to offer a critique of it. Those critiques fed into the development of her own distinctively formulated and culturally inflected version of an overdetermined Marxist political economy that shaped her conceptions of space as an integral and actively formative element of social and environmental relations. From this perspective, geographical space could not be separated from these wider relations and so it could not be understood and analysed purely in locational terms. Yet her recognition of the significance of idiographic Geography in revealing the distinctive complex qualities of place also helped inform the ways in which she worked with this approach thereby contributing so powerfully to her understanding of space and time and to her revelations of just how profoundly Geography and geography matter.

These are major achievements, the substantial reverberations of which continue to ripple way beyond Geography and into politics, social science and cultural studies. Founded in a grounded concern for place they reformulate both the idiographic contemplation of the regionally unique and the search for 'locational order'. They show how territorial difference is both distinctive and short-lived and how locational order is intimately related to wider social and economic relations which both shape and use geographical space and are shaped by it. But this profound reformulation of the nature of geographical space is far from all. The inseparability of Doreen's politics and Geography reformulated politics in active geographical terms focused on the power of place, on the relations of power exerted in and through space, and on the responsibility—not least through contributions to political analysis and the formulation of policy—for addressing uneven geographies of power. Like her powerful insights which transformed understanding of the nature and significance of geographical space, these concerns remain influentially ongoing in major publications —*Soundings* and *The Kilburn Manifesto*—in the creation of which she was both a prime mover and a subsequently powerful shaper.

The hope must be that contemporary politics will change as much and as progressively as the profound difference that she made to thought and practice in Geography and social science.

---

[208] P. Haggett, *Locational Analysis in Human Geography* (London, 1975), p. 2, emphasis in original

Acknowledgements
I am deeply indebted to all those who responded so positively to my invitation to discuss Doreen Massey's life and work. They all gave up many hours in order to meet me and/or to share their own writings. Their instant willingness to do so is itself a clear marker of Doreen's significance for their own lives and work. And, as is also abundantly clear from the large number of footnotes, their inputs form the core of everything written above. Radha Ray—Doreen's former PA at the Open University—responded instantly to all of my enquiries which spanned the eighteen months involved in the writing of this memoir. Linda McDowell and Trevor Barnes generously enabled me to have sight of the drafts of their chapters for *Critical Dialogues* and, equally generously, Gillian Rose sent me the link to her blog post on Doreen. And throughout the extended period of its compilation and writing Professors Felix Driver and Ron Johnston were never anything other than infinitely patient, supportive and encouraging. And, of course, I am very grateful to the British Academy for providing the opportunity to write this memoir.

Note: The identities of respondents so central to the writing of this Memoir have been anonymised.

Note on the author: Roger Lee is Emeritus Professor of Geography at Queen Mary University of London.

Anthony Barnes Atkinson

4 September 1944 – 1 January 2017

elected Fellow of the British Academy 1984

by

ANDREA BRANDOLINI

STEPHEN P. JENKINS

JOHN MICKLEWRIGHT

Biographical Memoirs of Fellows of the British Academy, XVII, 179–190
Posted 9 November 2018. © British Academy 2018.

TONY ATKINSON · *photo Dorothy Hahn*

Tony Atkinson died on 1 January 2017. He was one of the world's leading economists, an unrivalled scholar of inequality and poverty and among the founders of the modern study of public economics. His death was followed by an international outpouring of appreciation for both the man—one of the warmest, most generous, most decent of people—and his work, which forms part of the canon of social science.[1]

The career

> There are people who leave a lasting impact both on one's mind and one's heart. Tony Atkinson was one of them. (Nora Lustig)[2]

Tony was born the youngest of three sons in Caerleon, Wales, in 1944 and grew up in north Kent. He left school at seventeen and went to work for IBM, which he enjoyed, but within a year he moved to Hamburg—at the same time as the Beatles, he used to joke—to volunteer in a hospital in a deprived area for nine months, an experience which had an important impact on his outlook on life and its trajectory. He then went to Cambridge, to Churchill College, to study mathematics but had already decided to switch to economics after his first year. He had weekly supervisions in economics at first with Johannes Graaff and then with Frank Hahn. Not long ago, Tony wrote of Hahn, that '[t]he people who matter in one's intellectual development are those whose voices you hear in your head while doing your own research. Today I often hear Frank's voice as I work.'[3] In his third year he met Joe Stiglitz, then a Fulbright scholar, with whom he would publish one of his first papers and later (as we recount below) key contributions to the study of taxation as well as a ground-breaking graduate textbook on public economics. Tony went to lectures, from James Meade, Dick Goodwin, Jim Mirrlees and Peter Diamond among others, training him in the core curriculum of 'economic principles' and 'economic problems'. He noted approvingly to us in a conversation in 2014 that there was no distinction then between macroeconomics and microeconomics and he was taught economics as an integrated system. Every student had to have a copy of the UK national accounts (the 'Blue Book'). Tony was never to lose his interest in national income accounting, a subject later to fall out of the standard undergraduate economics curriculum. Forty years later he wrote an innovative report for the Office for National Statistics on the measurement of government output

[1] Many of these tributes have been collected together on Tony's website www.tony-atkinson.com (accessed 13 August 2018), which also lists his voluminous output of books and scientific papers.
[2] See the personal tributes from current and past members of ECINEQ's Executive Committee and Council at http://www.ecineq.org/image/TonyAtkinsonFinalVersion2.pdf (accessed 13 August 2018).
[3] This comment comes from Tony's contribution to an obituary of Frank Hahn in the 2012/13 *Economics Review* of the London School of Economics Department of Economics.

and productivity which has had considerable impact on the development of official statistics. He called for a 'principled approach'—a leitmotiv of his own contributions—focusing on measuring directly the output of the public sector: '[w]e cannot simply assume that outputs equal inputs in such a major part of the economy. To fail to measure the output would be to miss the essential complementarity between public services and private economic growth' (Atkinson, 2005, 182).

On graduating, Tony benefited from an exchange that Hahn had set up with the Massachusetts Institute of Technology and he went to spend a year in Cambridge, Massachusetts. He worked as a research assistant for Robert Solow and made lifelong connections with a range of young researchers, and was exposed to econometrics for the first time (he shared an office with Robert Hall, one of the original developers of the TSP econometric software package).

Tony did not go to the USA alone. He had met fellow student Judith Mandeville in Cambridge, England, and they married while still undergraduates. The partnership started with a shared interest: Tony and Judith met through a student society working for the benefit of the local community. They had three children, Richard, Sarah and Charles, and eight grandchildren (to whom Tony dedicated his final book: Atkinson, 2019). Tony and Judith celebrated their golden wedding anniversary in 2015.

Tony returned to the UK to a fellowship at St John's College, Cambridge. He related to us how he had sat at the kitchen table and planned a PhD thesis in the area of development economics, aiming to study dual economy models of agriculture and manufacturing, influenced by the work of Dale Mortensen. But his wide interests led him in other directions instead and, as was common at the time in the UK, he never did a PhD.

One of the few books that Tony had taken to the USA was *The Poor and the Poorest* by Brian Abel-Smith and Peter Townsend, published on Christmas Eve 1965, just after he and Judith had married. This was a landmark study of poverty in postwar Britain but Tony felt that it did not address what to do about the problem. The result was his own book, *Poverty in Britain and the Reform of Social Security*, which appeared in 1969, the first of more than forty books published over the next fifty years as author, co-author, editor or co-editor. Poverty was not a fashionable subject for economists, but James Meade was interested in the book's analysis and Tony described Brian Reddaway as being particularly encouraging and generous with his comments on a draft. Reddaway was a key early influence on Tony regarding the importance of taking data and data quality issues seriously, traits for which he became legendary.

At the same time as producing his empirical study of poverty and anti-poverty policy, Tony was working on the ideas set out in his enormously influential paper 'On the measurement of inequality', published in the *Journal of Economic Theory* in 1970—although he recalled to us that much of the paper itself was actually written while sitting on a boat in Lake Ohrid in the former Yugoslavia. We return to this paper below when we discuss some of Tony's fundamental contributions to social science. The point to be made here is that the 1970 paper is largely theoretical (albeit with huge practical implications described below). Throughout his career Tony produced both cutting-edge theory and the finest empirical analysis of data and policies, a rare combination in any social scientist.

In addition to his college fellowship in Cambridge, Tony was appointed subsequently to a university lectureship. But, in 1971 and together with his friend Christopher Bliss, he moved to the economics department at Essex, both of them being appointed to full professorships. This is especially noteworthy because Tony was only twenty-seven, much younger than the age at which people were appointed to chairs then, and even today. Tony has told us there was 'more freedom to develop things' at Essex, notably teaching. He probably felt uneasy with the intellectual atmosphere of Cambridge: 'As a student in Cambridge, England, in the early 1960s,' he wrote in the preface to his Graz Schumpeter Lectures, 'I was taught by a number of people who held strong views, views that they expressed with great fervour and combative style. Perhaps as a reaction, perhaps as a matter of personal temperament, I have always regarded my views about economics—in contrast to my moral principles— as tentative and open to revision as I learned more and acquired more evidence' (Atkinson, 2014a, x).

At Essex, Tony was able to teach public finance at the graduate level, as well as a course on the economics of inequality which resulted in his very successful 1975 book of the same name (a second edition appearing in 1983). He was at Essex between 1971 and 1975 before moving to head the then-named Department of Political Economy at University College London (UCL) between 1976 and 1979. A tribute to Tony after his death from the now-named Department of Economics at UCL includes the comment that 'it is typical of Tony that as head of department he was equally popular with the non-academic staff as with his professional colleagues', a feeling undoubtedly shared in all the other institutions in which he worked.

Tony moved to the London School of Economics and Political Science (LSE) in January 1980. He was Tooke Professor of Economic Science and Statistics in the Economics Department, where he taught enthusiastically at both graduate and undergraduate level. A key additional role was in the newly created Suntory-Toyota International Centre for Economics and Related Disciplines (STICERD) where he soon took over the chairmanship from the founder, Michio Morishima. Tony also brought support from the Social Science Research Council (soon to be renamed the Economic and Social Research Council: ESRC) for a programme of research called 'Taxation, incentives and the distribution of income', which he co-directed with Nick Stern and Mervyn King, then at Warwick and Birmingham respectively but who soon joined him at the LSE. This was one of the first times that the ESRC had funded a long multi-project programme rather than individual projects. The programme lasted twelve years. Nick Stern has said 'they were special years' and he has emphasised Tony's role in building the programme and building STICERD.[4]

In 1992 Tony returned to Cambridge and Churchill College, where he had started his university career and where he had met Judith. He said he had gone back thinking he would eventually retire there. But he did not feel at home and left after two years when the opportunity arose to be the Warden of Nuffield College, Oxford, a post he held from 1994 to 2005. Tony had also been considered for the Presidency of the European University Institute in Florence, a role that would have suited him very well although he would probably have found less time for his research and teaching, which continued to flourish at Nuffield. On stepping down from the Wardenship, and after a year at the Paris School of Economics, Tony had a chair in Oxford for three years before, to his chagrin, being obliged to retire from the post at age sixty-five under the terms of the existing legislation. He continued to be an Honorary Fellow at Nuffield and part-time Centennial Professor at the LSE until his death.

Tony's public service and wider professional activities were wide-ranging and numerous and we list only a few here. In the 1970s he co-founded and edited a leading journal (*The Journal of Public Economics*, discussed further below) and he was a member of the Royal Commission on the Distribution of Income and Wealth. In the 1980s he was a specialist adviser to two House of Commons Select Committees. In the 1990s he was a member of the Conseil d'Analyse Economique, advising the French Prime Minister Lionel Jospin and he became a trustee of the Nuffield Foundation. From 2011 until his death he was president of the Luxembourg Income Study (the most comprehensive international database for cross-country comparisons of income and wealth). He was president at different times of all the leading UK and international economics societies: the Royal Economic Society (1995–98), the European Economic Association (1989), the Econometric Society (1988), and the International Economic Association (1989–92). Honours bestowed on him included some twenty honorary doctorates, his fellowships of the Econometric Society (1974) and of the British Academy (1984),[5] foreign honorary memberships of the American Economic Association (1985) and American Academy of Arts and Sciences (1994), and the Dan David Prize (2016, joint with François Bourguignon and James Heckman). Tony was knighted in 2000 for his services to economics and made a Chevalier de la Légion d'Honneur in 2001.

---

[4] See Stern's introduction to his interview with Tony (Atkinson and Stern 2017).
[5] Tony Atkinson served as a Vice-President of the British Academy, 1988–90.

The work

> We have become too specialized, and people define themselves as being specialized economists [a labour economist, an IO—industrial organisation—economist etc.], whereas I just think of myself as an economist. (Tony speaking to Nick Stern in 2016—see Atkinson and Stern, 2017, 15)

Tony's research activities were wide-ranging, far broader than most other economists of his generation or of later cohorts. This breadth is highlighted by our article with fourteen other authors that reviews 'Tony Atkinson and his legacy'. As we write in the introduction, 'Tony was an economist in the classical sense, rejecting any label of a sub-field on his interests and expertise' (Aaberge et al., 2017, 421). For a proper appreciation of the range and depth of Tony's work and its influence, we point readers to that article and to another extended assessment of his contribution by Jenkins (2017). Here we aim only to give a flavour, focusing on a few of his most important writings.

Tony's 1970 *Journal of Economic Theory* paper radically changed the way that economists think about the measurement of inequality, providing the starting point for the modern analysis of the subject. The article made three crucial contributions. First, and in line with his firmly held view that economics is a moral science, he made clear that all measures of income inequality are underpinned by normative assumptions, which can be encapsulated in what economists call a social welfare function. There is no purely 'statistical' index of inequality (different indices give more weight to the top, middle or the bottom of the distribution of income, for example). Second, Tony showed that it is straightforward to check whether one distribution of income is more, or less, unequal than another (e.g. the UK in 2010 compared with the UK in 2000) according to all standard inequality indices regardless of the normative weights built into them: there is a simple comparison based on easily constructed graphs known as Lorenz curves. Third, he introduced a new family of inequality measures that makes different views about distributional justice explicit through a parameter capturing the 'inequality aversion' of the measurer—the now widely used Atkinson class of inequality indices.

Tony's insights stimulated a whole new field of research, and he accompanied it with original thinking about the reasons why inequality was changing within countries and between rich countries around the world, often rejecting conventional assumptions and models. The result was that Tony succeeded in '[b]ringing income distribution in from the cold', to use the title of his Presidential Address to the Royal Economic Society (Atkinson, 1997). The analysis of inequality is now more within the mainstream of economic analysis and perhaps even fashionable. Tony's insights on the measurement of inequality also cross-fertilised work on how to measure poverty. His Presidential Address to the Econometric Society, 'On the measurement of poverty', was a fundamental contribution to this trend and has been very influential (Atkinson, 1987). Among Tony's Google Scholar citations (as at May 2018), the 1970 paper about inequality measurement tops the list but his 1987 paper on poverty measurement is number five.

Advancing the analysis of inequality and poverty is not only a matter of theory. Proper attention needs to be paid to the data used: their characteristics and fitness-for-purpose, and their limitations whenever they are only a proxy for the ideal theoretical concept. We have already referred to Tony's 1969 book on poverty as exhibiting this care. Throughout his career, in paper after paper, in monographs, in reports and in edited volumes, Tony demonstrated the same qualities—in work on the UK, on Europe, on the advanced industrialised countries of the Organisation for Economic Co-operation and Development, and at the global level. We pick out just two recent examples, both written after the onset of the illness that led to his death.

Our first example is Tony's 2015 magnum opus *Inequality: What Can Be Done?* (Atkinson, 2015), now translated into some fifteen other languages. An early version of the book's ideas was published in the *British Journal of Sociology* (Atkinson, 2014b); Tony was no hidebound economist—he was always appreciative of the contributions from other disciplines. He was very appreciative of the work of the French economist Thomas Piketty, whom he had mentored and collaborated with on the measurement of trends in top incomes (e.g. Atkinson and Piketty, 2007, 2010). Piketty's bestselling 2014 book, *Capital in the Twenty-First Century*, did much to draw popular attention to long-run trends in income inequality in the Western world. But Tony wanted to go further and show how inequality could actually be reduced in practice. Tony's book, as ever elegant, clearly written, full of knowledge of historical precedent and earlier literature, offers fifteen proposals for altering the distribution of income, through changes in the labour market and capital market as well as through taxation and social benefits.

Our second example is Tony's work on global poverty, specifically the report he wrote for the World Bank advising on its measurement (Atkinson, 2017) and a book building on his report which is to be published fifty years after his first (Atkinson, 2019). In his report, Tony dissected the World Bank's approach to measuring 'extreme poverty', assessed using the $1.90-per-day poverty line fixed in terms of international dollars (US dollars adjusted for differences in purchasing power across countries). This line is used to formulate a key target in the first of the United Nations' Sustainable Development Goals to which world leaders committed themselves in 2015. Tony's report highlights various problems with the World Bank's approach, though he recommended maintaining the $1.90 poverty line because it 'has acquired an independent political status' (Atkinson, 2017, 30). Tony observes that a truly global approach must also cover high income countries, and he recommends that a broader set of complementary indicators be used. The World Bank has committed to implement many of Tony's recommendations.

In contrast with the 2017 report, in his 2019 book Tony starts from first principles, discussing how to measure poverty across the world, in both rich and poor countries. These foundational principles are translated into concrete measures, and Tony then analyses the available data to which the measures can be applied, dealing with measurement based on both low income or expenditure (as in the World Bank's focus) and on non-monetary indicators of deprivation that have gained increasing attention in recent years in countries at all levels of development. Critically, the book bridges international organisations' measurement of poverty with national analyses produced within countries, an important innovation.

To this point we have focused on Tony's work on inequality and poverty. But he also made huge contributions to the study of public economics. This 'is an exciting topic. It deals with key issues facing governments all round the world. The subject matter—taxes, government spending, and the national debt—is very much in the minds of citizens and their elected representatives. What is more, it draws on the whole range of economics. To understand public policy, it is necessary to consider how households and firms make decisions, how they interact in a market economy, and how the economy develops over time' (Atkinson and Stiglitz, 2015, xi). Tony's co-author from the first decades of his career, Nobel laureate Joe Stiglitz, has written that 'Tony's work in public economics changed forever the field ... so much so that it is hard to even recognise the subject as it previously existed' (Aaberge et al., 2017, 428).

Tony's contributions came in three ways. First, he produced seminal papers, many of which underlined the positive role that the state can play in the economy, especially the welfare state, or which demonstrated that existing views of the state's role were too simple. In this connection, we mention only two journal articles written early in his career jointly with Stiglitz, one on the theory of optimal taxation—how best to design taxes to raise a given amount of revenue in a world where individuals differ (economic models too often still assume that they do not)—and another on where the balance between direct and indirect taxation needs to be considered (Atkinson and Stiglitz, 1972, 1976). Second, again with Stiglitz, Tony wrote a landmark graduate textbook, *Lectures on Public Economics*, published in 1980, that for decades has been essential reading for master and doctoral students around the world. It was re-published in 2015 at a time when a second-hand copy could fetch a three-figure sum (the definition of public economics above comes from the introduction to this second edition). This was one of those few books you simply *had* to read when entering the world of graduate study in economics. Third, Tony founded in 1971 and then co-edited for the next quarter century a new journal, the *Journal of Public Economics*, which quickly established itself as the leading journal in the field.

Tony was no ivory tower academic in his work on policy. He was also heavily involved as a policy advisor nationally and internationally, and distinctive among economists worldwide in also making huge contributions to the development of the statistical monitoring systems required to accompany effective policy analysis. For example, the European Union's indicators for monitoring social inclusion are largely due to Tony's work with collaborators (see inter alia Atkinson et al., 2002).

No account of Tony's work could finish without explicit reference to his teaching. He was a dedicated teacher, both in the lecture theatre and in one-to-one graduate student supervision. His lecturing dovetailed with his writing. *Lectures on Public Economics* and *The Economics of Inequality*, products of his teaching, are numbers two and three in his Google Scholar citations. We do not know precisely how many PhD students Tony formally supervised but, as of 2007, it was at least sixty and in addition there are many other younger scholars whom he influenced directly through his collaboration on joint research projects. Tony was a great believer in collaborative work and he nurtured many careers as a result—he placed great store on bringing on the next generation of researchers. He was patient, encouraging and always available. His guidance was 'ever present but never irksome' as one of us wrote in the acknowledgements to his PhD thesis.

The man

> Tony was one of the most decent, kindest people I have ever met. (Jane Dickson, his secretary for many years at the LSE)[6]

In writing about Tony's career and work we have already implied much about the man—someone held in the highest regard for his human qualities as well as for his intellect and output. He brought out the best in people, always able to engage with people as he found them and to listen with respect to their point of view.

Although this memoir is mostly about Tony and economics, we must emphasise that he had many interests outside economics and his work. He was a passionate sailor and owned a succession of different boats. Much of Tony's sailing was done single-handed in the tidal waters off the Essex coast and the reliance on his physical abilities and judgement that this required brought him a great deal of satisfaction. He shared with Judith a love of literature and of walking—together they walked the Pennine Way, part of the pilgrim route to Santiago de Compostela, and across Normandy, an area of France to which Tony was deeply attached.

---

[6]See the LSE's virtual memory wall for Tony at http://sticerd.lse.ac.uk/atkinson/default.asp (accessed 13 August 2018).

To some people, Tony was an archetypal Englishman, with his modest and unassuming manner. Yet he was an internationalist in outlook, including being a passionate European. His engagement with people around the world not only gave him enjoyment but had substantial effects on the work that he did and on the people he met face to face, whom he lectured to, and who read his articles and books. Tony acknowledged the many social challenges of the twenty-first century, including population ageing, climate change and global imbalances. But rather than being pessimistic about the prospects for social progress, Tony made a convincing and profoundly inspirational case that there are 'grounds for optimism' and that 'the solutions to these problems lie in our own hands' (Atkinson, 2015, 308). Tony leaves a huge legacy of work that is stimulating new generations of researchers and there are many people worldwide who remember fondly how they were touched by his humanity. In these senses, Tony lives on.

Acknowledgements
We have drawn on a conversation with Tony in Oxford in November 2014 during which he discussed with us his early career and especially his time in Cambridge.

Note on the authors: Andrea Brandolini is head of the Statistical Analysis Directorate at Banca d'Italia. Stephen P. Jenkins is Professor of Economic and Social Policy at the London School of Economics and Political Science. John Micklewright is Professor Emeritus in the Department of Social Science in the University College London Institute of Education.

References

Aaberge R., Bourguignon, F., Brandolini, A., Ferreira, F. G., Gornick, J. C., Hills, J., Jäntti, M., Jenkins, S. P., Marlier, E., Micklewright, J., Nolan, B., Piketty, T., Radermacher, W. J., Smeeding, T. M., Stern, N. H., Stiglitz., J. E. and Sutherland, H. (2017). 'Tony Atkinson and his legacy', *Review of Income and Wealth*, 63, 412–44.

Abel-Smith, B. and Townsend, P. (1965). *The Poor and the Poorest*. London: G. Bell and Sons.

Atkinson, A. B. (1969). *Poverty in Britain and the Reform of Social Security*. Cambridge: Cambridge University Press.

Atkinson, A. B. (1970). 'On the measurement of inequality', *Journal of Economic Theory*, 2, 244–63.

Atkinson, A. B. (1975). *The Economics of Inequality*. Oxford: Clarendon Press (2nd edn 1983).

Atkinson, A. B. (1987). 'On the measurement of poverty', *Econometrica*, 55, 749–64.

Atkinson, A. B. (1997). 'Bringing income distribution in from the cold', *Economic Journal*, 107, 297–321.

Atkinson, A. B. (2005). *Measurement of Government Output and Productivity for the National Accounts*. Basingstoke: Palgrave Macmillan.

Atkinson, A. B. (2014a). 'After Piketty?', *British Journal of Sociology*, 65, 619–38.

Atkinson, A. B. (2014b). *Public Economics in an Age of Austerity*. London: Routledge.

Atkinson, A. B. (2015). *Inequality: What Can Be Done?* Cambridge, MA: Harvard University Press.

Atkinson, A. B. (2017). *Monitoring Global Poverty. Report of the Commission on Global Poverty.* Washington DC: The World Bank.

Atkinson, A. B. (2019). *Measuring Poverty Around the World.* Princeton, NJ: Princeton University Press.

Atkinson, A. B., Cantillon, B., Marlier, E. and Nolan, B. (2002). *Social Indicators: the EU and Social Inclusion.* Oxford: Oxford University Press.

Atkinson, A. B. and Piketty, T. (eds.) (2007). *Top Incomes over the Twentieth Century: a Contrast between Continental European and English-Speaking Countries.* Oxford: Oxford University Press.

Atkinson, A. B. and Piketty, T. (eds.) (2010). *Top Incomes: a Global Perspective.* Oxford: Oxford University Press.

Atkinson, A. B. and Stern, N. H. (2017). 'Tony Atkinson on poverty, inequality, and public policy: the work and life of a great economist', *Annual Review of Economics*, 9, 1–20.

Atkinson, A. B. and Stiglitz, J. E. (1972). 'The structure of indirect taxation and economic efficiency', *Journal of Public Economics*, 1, 97–119.

Atkinson, A. B. and Stiglitz, J. E. (1976). 'The design of tax structure: direct versus indirect taxation', *Journal of Public Economics*, 6, 55–75.

Atkinson, A. B. and Stiglitz, J. E. (1980). *Lectures on Public Economics.* Maidenhead: McGraw-Hill. (Updated edition: Princeton, NJ: Princeton University Press, 2015.)

Jenkins, S. P. (2017). 'Anthony B. Atkinson (1944–)', chapter 52 in R. Cord (ed.), *The Palgrave Companion to Cambridge Economics.* Basingstoke: Palgrave Macmillan, pp. 1051–74.

Piketty, T. (2014). *Capital in the Twenty-First Century.* Cambridge, MA: Harvard University Press.

Michael Anthony Eardley Dummett

27 June 1925 – 27 December 2011

elected Fellow of the British Academy 1968

resigned 1984

re-elected Fellow of the British Academy 1995

by

DANIEL ISAACSON

IAN RUMFITT
Fellow of the Academy

Biographical Memoirs of Fellows of the British Academy, XVII, 191–228
Posted 21 November 2018. © British Academy 2018.

MICHAEL DUMMETT

Throughout the second half of the twentieth century, Michael Dummett was a powerful figure in British philosophy and in later years its most distinguished and authoritative practitioner, with tremendous international standing. His work spanned philosophy of mathematics, formal logic, philosophy of language, history of philosophy, and metaphysics. He also played an important role in combatting racism in Britain, and when he was knighted, in the 1999 New Year's Honours, it was 'for services to Philosophy and to Racial Justice'.

Biography

Early life and education

Michael Dummett was born at 56 York Terrace, London, his parents' home, on 27 June 1925, and died on 27 December 2011 at 54 Park Town, Oxford, the home where he and his wife Ann had lived since 1957 and brought up their children. He was the only child of his parents George Herbert Dummett (1880–1970), a silk merchant, who also later dealt in rayon, and Mabel Iris née Eardley-Wilmot (1893–1980), whose father, Sir Sainthill Eardley-Wilmot, had been Inspector-General of Indian Forests, and after whom Michael Dummett was given Eardley as his middle name. Dummett's father had two sons and a daughter by a previous marriage.

At the age of ten Dummett was sent as a boarder to a preparatory school, Sandroyd, in Cobham, Surrey. In September 1939, at the onset of the Second World War, he began his secondary education at Winchester College, having come top of the election roll for Scholars. After a compulsory year on the 'classics ladder', he opted for science, but was 'deeply disappointed' by it and switched to history.[1] In 1943 he obtained a history scholarship to Christ Church, Oxford, but—now eighteen and with the war still raging—went instead into the Royal Artillery, under which auspices he was sent on a six-month 'short course' at Edinburgh University. There he contacted the Catholic Chaplaincy and underwent instruction by the Chaplain, Father Ivo Thomas, and was received into the Roman Catholic Church on 10 February 1944. Dummett took the confirmation name Anthony, after St Anthony of Padua, and used it as a middle name thereafter, in addition to the middle name given to him by his parents. A child of irreligiously Anglican upbringing, and a declared atheist at fourteen, the deep religious faith of his conversion remained central throughout the rest of his life, though not always without struggle.[2]

[1] M. A. E. Dummett, 'Intellectual autobiography', in R. L. Auxier and L. E. Hahn (eds.), *The Philosophy of Michael Dummett*, Library of Living Philosophers (Chicago, IL, 2007), p. 4.
[2] Ibid., pp. 5–6.

Dummett was then sent for six weeks of army basic training, after which he went on a six-month course at the ISSIS (Inter-Service Special Intelligence School) in Bedford to be taught to read and translate Japanese, and then to the Wireless Experimental Centre outside Delhi to translate intercepted Japanese messages. (The Wireless Experimental Centre was one of two overseas outposts of Station X in the signals analysis centre at Bletchley Park, and Dummett is recorded in the online Roll of Honour of Bletchley Park for this work.[3]) When the war with Japan ended, he was sent to Malaya as part of Field Security. He wrote that 'it must have been in Malaya that a passionate hatred of racism was first born in me. I learned of the means by which the British masters of pre-war colonial Malaya had maintained and acted out the myth of white racial superiority',[4] though Michael Screech, who was on the Bedford course and at the Wireless Centre with Dummett, remembered him expressing anger about racism already at that time. Dummett was by then a heavy smoker, as he remained throughout his life, and Screech recalled that tapping the end of a cigarette many times before lighting it came to be called 'dummetting' by those around him.

Dummett was demobilised in 1947, with the rank of sergeant,[5] and took up the scholarship he had been awarded at Christ Church, Oxford during the war. A letter from Harold Walker, Dummett's history master at Winchester (who had played a key role in enabling Dummett to make history his main subject,[6] it being assumed in Winchester at that time that its most brilliant students, of whom Dummett was clearly one, should do either classics, or sciences and mathematics) to Nowell Myres, Winchester's main contact in Christ Church, on 21 July 1947, gives a snapshot of Dummett at a crossroads:

> Michael Dummett, who is coming up to Ch. Ch. as a History Scholar in October, was staying with me this weekend. He is just back from Malaya where he has been a ser-geant in the Intelligence, the Army having taught him Japanese. He asked my advice as between History & PPE, and I recommended him to do History [...] He is a very able boy—actually he's 22 now! – & was senior on the roll. But he's the last person in the world to do PPE. He has always needed steadying rather than stimulating,—he

---

[3] https://bletchleypark.org.uk/roll-of-honour/2681 (accessed 6 September 2018).
[4] Dummett, 'Intellectual autobiography', p. 8.
[5] 'Dummett, Sir Michael (Anthony Eardley)' entry in *Who's Who*, http://www.ukwhoswho.com/view/10.1093/ww/9780199540891.001.0001/ww-9780199540884-e-14263 (accessed 13 September 2018).
[6] Dummett refers to Harold Walker as 'an inspired teacher' (Dummett, 'Intellectual autobiography', p. 4); for an account of Harold Walker's impact on a pupil at Winchester who went on to a distinguished career as a historian, see the British Academy Biographical Memoir for Nicholas Brooks: B. Crawford, S. Keynes and J. Nelson, 'Nicholas Brooks', *Biographical Memoirs of Fellows of the British Academy*, 15 (London, 2016), pp. 23 and 28–9.

stimulates himself! Even at 15, before I rescued him from the Science Side, he was apt to be writing learned papers on Chinese Art to deliver to College societies, or articles against the Public Schools for Picture Post, to the neglect of his other work. Since then he has progressed from Scepticism through Medieval Mysticism to the Roman Church, where he may or may not remain.

Dummett opted for the Honour School of Philosophy, Politics, and Economics (PPE) in part because he felt that after four years in the army he had forgotten much of the history he had learned. He was 'soon captivated by philosophy'.[7] His philosophy tutors were Michael Foster, Anthony Flew, J. O. Urmson and David Pears, who was appointed a research lecturer at Christ Church in 1948, and who later (in 1973) wrote of Dummett, 'I first met him when he was an undergraduate at Christ Church and was impressed by his penetration. In discussion he would quickly put the superficial issues on one side, and go directly to the fundamental ones.' Dummett was also sent for tutorials to Elizabeth Anscombe at Somerville, whose commitment to Wittgenstein greatly influenced Dummett at that time. For his Finals, in 1950, Dummett chose to do a paper established by J. L. Austin called 'The Origins of Modern Epistemology', available for the first time that year. Candidates were expected to study four texts from a list of seven, one of which was Frege's *The Foundations of Arithmetic*, translated from the German by Austin for this purpose. Instruction was provided by William Kneale and Friedrich Waismann in a class they gave on Frege's *Grundlagen* in Hilary Term 1950.[8] Dummett later wrote of Frege's *Die Grundlagen der Arithmetik*, 'I thought, and still think, that it was the most brilliant piece of philosophical writing of its length ever penned.'[9] Dummett's ensuing work on Frege transformed understanding of this seminal figure's logic and philosophy.

Early career

After taking First Class Honours in PPE Finals in the summer of 1950, Dummett was appointed to a one-year Assistant Lectureship in Philosophy at the University of Birmingham. That October he sat the fellowship examination at All Souls College, Oxford, and was elected, with immediate effect, but nonetheless fulfilled his commitment to Birmingham, rushing back to Oxford during term to pernoctate as required by All Souls.

The first project Dummett set himself as a Prize Fellow at All Souls was to read all the published work of Frege, most of which at that time had been neither translated

---

[7] Dummett, 'Intellectual autobiography', p. 9.
[8] *Oxford University Gazette,* 14 December 1949, 27.
[9] Dummett, 'Intellectual autobiography', p. 9.

nor republished. He also visited the Frege archive in Münster to study what survived of Frege's *Nachlass*. Despite his passion for Frege, Dummett began his philosophical career thinking of himself as a follower of Wittgenstein, arising from the impact of the arrival in Oxford during his last year as an undergraduate of typescripts of *The Blue and Brown Books* and of notes of Wittgenstein's classes on philosophy of mathematics, and from his philosophical contact and ensuing friendship with Elizabeth Anscombe. By 1960 he no longer considered himself a Wittgensteinian.[10]

On 31 December 1951, in his second year as a Prize Fellow, Dummett married Ann Chesney (1930–2012), who had taken Finals in History from Somerville College that summer. She was the daughter of the actor Arthur William Chesney. Fifty years later Dummett wrote of Ann, 'she has been my constant support and delight throughout my life'.[11] They had seven children, four sons and three daughters, of whom two, a son and daughter, died in infancy. To support his growing family, Dummett took on a great deal of undergraduate teaching for other colleges, 'since All Souls then paid its Fellows no marriage allowance, housing allowance, or children's allowance, and, unlike other colleges, had no houses to rent to Fellows on preferential terms'. Dummett 'once complained to John Sparrow, then Warden, that the College had houses for servants but not for Fellows: he replied that it was difficult to get servants'.[12]

Early in his All Souls Fellowship, Dummett had the idea of doing a second BA in Mathematics, but Humphrey Sumner, the then Warden, refused permission, on the grounds that it would disgrace the College if he failed to obtain a First, and he settled for some tutorials with John Hammersley, an applied mathematician in Oxford, later a fellow of Trinity College. Dummett was awarded a Harkness Fellowship to spend the academic year 1955–56 at the University of California, Berkeley, studying logic and mathematics. Ann and their two young children managed to join him there for seven months, on his very limited stipend. He learned a great deal from Leon Henkin, Raphael Robinson, John Myhill, Paul Halmos and others (but not Alfred Tarski, who was away that year). He also at that time came to know Donald Davidson, who was then teaching at Stanford, and they remained firm friends and philosophical interlocutors to the end of Davidson's life.

Mid-career and anti-racism

While in Berkeley Dummett became closely involved with the American civil rights movement. He noted later that 'at that time the United States was the most racist

[10] M. A. E. Dummett, *Truth and Other Enigmas* (London, 1978), p. xii.
[11] Dummett, 'Intellectual autobiography', p. 10.
[12] Ibid., p. 11.

country in the world after South Africa'.[13] He and Ann joined the National Association for the Advancement of Colored People and attended a rally in San Francisco addressed by Dr Martin Luther King. Part of the duty of a Harkness Fellow was to travel in the United States, and after Ann and the children returned to England, Dummett devoted himself to visiting black Americans during that summer of 1956. He travelled to Montgomery, Alabama, where the boycott organised by Martin Luther King to force, by peaceful means, repeal of the law segregating blacks on the city's bus system was in progress. He there met Dr King, whom he admired greatly.

In 1957 Dummett was elected to a further seven years as a Fellow of All Souls. He was also, in November 1957, offered appointment as Assistant Professor in the Philosophy Department at the University of California, Berkeley, for which he had not applied; the Department was keen to hire him having seen his exceptional qualities during his year at Berkeley as a Harkness Fellow. After considerable correspondence and agonising, Dummett accepted the offer, in April 1958, on the basis that he would start in September 1959, but in November 1958 he withdrew his acceptance, he and Ann having come to the conclusion that they should bring up their children in England rather than America: 'We neither of us wanted our children to grow up in an environment alien to us which we did not truly understand.'[14]

During Trinity Term 1958 (March–June) Dummett went on his own to the University College of Ghana in Legon as a Visiting Lecturer.[15] In Legon he taught epistemology,[16] and gave a seminar on the philosophy of time, one of his developing interests. He also worked on his ideas for topological models of modal logic during this period, which he had begun earlier with E. J. Lemmon, and which came close to the notion of a Kripke model, without quite reaching it. He became ill, however, and did not return to the topic: 'possibly, if I had, what are now called Kripke models might have been called Dummett–Lemmon models, though I doubt if we should have

---

[13] Ibid.

[14] Ibid., p. 16, provides a somewhat condensed account of this episode.

[15] The invitation to Legon was at the behest of William Abraham, a Ghanaian who had arrived in Oxford in Michaelmas Term 1957 as a BPhil. student directly after having taken his BA in Philosophy in Legon, and who came to know Dummett through attending his lectures and classes. When Dummett welcomed Abraham's suggestion that he visit Legon, the head of the department, Daniel Taylor, immediately arranged the invitation. When Abraham completed the BPhil. in 1959, Dummett encouraged him to sit the All Souls Prize Fellowship exam, and Abraham was elected, the first and so far only African Prize Fellow of All Souls. In correspondence to do with the writing of this memoir, Abraham noted that, 'One other thing for which I am grateful to Michael is that he pointed me at the Catholic Church. He regularly gave me *Blackfriars* to read. I was received into the Catholic Church in March 1967.'

[16] Letter from William Abraham 8 May 2018, based on enquiry to Kwasi Wirudu.

thought what to do about the quantifiers'.[17] It was while in Ghana that Dummett sent his acceptance of the Berkeley offer, by cable and then aerogramme.

In 1950 P. F. Strawson published 'On referring', rejecting Russell's theory of descriptions (1905: cited by Frank Ramsey as 'that paradigm of philosophy'), on which reference failure renders a sentence false, in favour of a doctrine of presupposition, on which such sentences are neither true nor false.[18] During the late 1950s Dummett began to explore the implications of those ideas for the notions of truth and logical validity, of which he gave the following account in the Preface to his 1978 collection of previously published papers *Truth and Other Enigmas*:

> Interest in the doctrine of presupposition had led me to an interest in the concept of truth: and this, in turn, led me to an interest in the question how, if at all, it is possible to criticise or question fundamental logical laws that are generally accepted. These are interests that have remained preoccupations throughout my philosophical career. Their first fruit was a book called *The Law of Excluded Middle*, based on lectures that I had given in Oxford, that I submitted, I think in 1958, to the Oxford University Press and that was accepted by it on the advice of the late Professor Austin, one of the delegates of the Press. Austin was kind enough to recommend publication of the book [...] He had, however, reservations about my literary style, and required as a condition of publication, substantial stylistic emendation. At the time, of course, I found this galling, but could do nothing but agree; but, as I engaged in the laborious process of trying to comply, I became more and more dissatisfied with the content of the book, and never resubmitted it.[19] In a sense I have been trying to rewrite the book ever since.[20]

In his 'Intellectual autobiography', Dummett sharpens this earlier account by stating, 'I should be ashamed of it now if it had been published.'[21] He there goes on to say, 'This led me to study intuitionistic logic and the intuitionist philosophy of mathematics, to which I felt strongly drawn.' Also at this time he had been writing his massive review article, which he published in 1959, of Wittgenstein's *Remarks on the Foundations of Mathematics*, published in 1956.[22]

The confluence of these ideas came together in his paper 'Truth', published in 1959, a seminal work and his single most important paper.[23] It contained within it the

---

[17] Dummett, 'Intellectual autobiography', p. 16.

[18] P. F. Strawson, 'On referring', *Mind*, 59 (1950), 329.

[19] No copy of this manuscript is known to have survived.

[20] Dummett, *Truth and Other Enigmas*, pp. xix–xx.

[21] Dummett, 'Intellectual autobiography', p. 15.

[22] M. A. E. Dummett, 'Wittgenstein's philosophy of mathematics', *Philosophical Review*, 68 (1959), 324–48.

[23] M. A. E. Dummett, 'Truth', *Proceedings of the Aristotelian Society*, n.s. 59 (1959), 141–62. The paper was presented at a meeting of the Aristotelian Society on 16 February 1959. William Abraham reports

seeds of a great deal of his later philosophy. It adumbrated the opposition between realism and anti-realism, as Dummett characterised these positions, in terms of bivalence and the law of excluded middle, and surveyed a variety of contexts, both mathematical and non-mathematical, in which this opposition arises. A connection between these considerations and Wittgenstein's dictum that meaning is use was sketched. It also contains a somewhat offhand rejection of Strawson's idea that reference failure results in truth value gaps ('It is thus prima facie senseless to say of any statement that in such-and-such a state of affairs it would be neither true nor false'[24]). Late in the paper Dummett arrives at the idea that mathematical intuitionism is the paradigm for the anti-realism he there adumbrates: 'What I have done here is to transfer to ordinary statements what the intuitionists say about mathematical statements.'[25] This heady mixture of ideas took decades to explore, and led to new understanding of the nature of logic and meaning.

In 1962 Dummett applied for and was appointed to the Oxford University Readership in Philosophy of Mathematics, in succession to Hao Wang (who had succeeded Friedrich Waismann, the first incumbent of the post), which he held in conjunction with a Fellowship of All Souls. Between 1960 and 1966 he was regularly a visiting professor in the Philosophy Department at Stanford for the summer quarter (in part to earn money so that he could take his family on holiday). During one of those visiting appointments, in 1964, he gave a course of lectures as a preliminary version of a book he hoped to write surveying every variety of realism or denial of realism, the first of his attempts to rewrite the book that Oxford University Press had accepted for publication, subject to stylistic revision, in 1958, but when he returned to Oxford that summer he and Ann decided 'that the time had come for organised resistance to the swelling racism in England',[26] and he put this project on hold, along with a book on Frege he had been planning.

For the next four years, Dummett devoted every moment he could spare to the fight against racism, while fulfilling his heavy teaching commitments. He and Ann were deeply engaged both in organisational activity to combat racism as a trend in British government and society, and in work on behalf of individuals threatened by racist policies and attitudes. This latter included intervening to stop persons of colour from being deported back to the country from which they were fleeing as they

---

that at Dummett's invitation, he and Gillian Romney, another BPhil. student, attended this meeting, at which A. J. Ayer presided (presumably deputising for Karl Popper, who was President of the Aristotelian Society in that year). Attendance at this meeting was sparse, and there was little comprehension of the paper.

[24] Ibid., p. 150.
[25] Ibid., p. 160.
[26] Dummett, 'Intellectual autobiography', p. 19.

attempted to enter the UK. A telephone call at any hour of the day or night would alert Dummett to such a case, and transform him from philosopher to activist, telephoning the Chief Immigration Officer to obtain a stay of immediate deportation, then dashing to the airport to argue the case, often successfully. Dummett's organisational work against racism included a role in founding the Oxford Committee for Racial Integration, participation in the turbulent and ultimately self-destructive Campaign Against Racial Discrimination, and playing a key role in founding the Joint Council for the Welfare of Immigrants (JCWI), which continues to do important work to the present day. He chaired its inaugural meeting at the Dominion Theatre in Southall in September 1967, and maintained his association with it to end of his life. Its website contains an obituary of Dummett by Habib Rahman, its then Chief Executive, who remembered Dummett as 'an extremely compassionate person and a fierce opponent of racism in our society', and declared, 'He will be sorely missed and fondly remembered by us for his uncompromising struggle for equal rights for migrants and refugees in Britain.'[27] Ann is also memorialised by an obituary on the JCWI website. Dummett later described this time as 'the most exhausting period of my life'.[28] One could say that he had anti-racism in his genes, from the fact (discovered, to his great delight, by his daughter Suzie) that Sir John Eardley Eardley-Wilmot, grandfather of his mother's father, had campaigned for the abolition of slavery, and is among those depicted in the painting by Benjamin Robert Haydon, hanging in the National Portrait Gallery, of the Anti-Slavery Society convention of 1840.

During this period, while on holiday in France with his family, Dummett came across, by chance, a pack of Tarot cards *avec règles du jeu* in a shop, which he bought for entertainment during the holiday, and the game was enjoyed in the family. Back in England, he 'came across an Austrian pack, also with rules: the game was very different, although plainly related. I wanted to discover how the game was played in other countries, and wrote to card-game experts to ask, but none of them could tell me. I then embarked on my own enquiries.'[29] Thus began a passionate side interest in Tarot, in the cards themselves, and in the games played with them: 'It may seem odd that I could pursue a new interest in the midst of involvement in the struggle against racism. It was a solace. It provided difficult intellectual problems whose solution, unlike those of philosophical ones, had no serious import: it relieved the anxiety that always accompanied thinking about the racial situation or the problems of individuals entangled in it.'[30]

---

[27] https://www.jcwi.org.uk/2012/01/06/sir-michael-dummett (accessed 6 September 2018).
[28] Dummett, 'Intellectual autobiography', p. 21.
[29] Ibid., p. 23.
[30] Ibid.

While maximally committed to anti-racism in this time, Dummett fulfilled all his obligations as a teacher, in which he was inspiring, and also played an important role in establishing mathematical logic within Oxford University. This resulted in the creation in 1965 of a University Lectureship in Mathematical Logic associated with a Fellowship at All Souls, to which John Crossley was appointed. Dummett then, together with Crossley, did a great deal of the work in creating a new Oxford undergraduate degree in Mathematics and Philosophy, which the Kneale Report in 1966 had called upon the University to establish (along with a joint school of Physics and Philosophy). Dummett gave the lead to colleagues on the committee that was set up by the Faculty Board of Literae Humaniores, of which Philosophy was then a Sub-Faculty, and the Mathematics Faculty Board, to design the new course: 'As bridge subjects we included in the curriculum the philosophy of mathematics and a very large component of mathematical logic, including an optional paper on intuitionism.'[31] The teaching of the bridge subjects fell largely to Dummett and Crossley, when the joint school got underway, and they found themselves having to give twice the number of tutorials and lectures required by their conditions of appointment, a situation which continued with Crossley's successor, Robin Gandy, who came to Oxford in 1969 as Reader in Mathematical Logic This situation was finally somewhat alleviated with the establishment, in 1971, of a Professorship in Mathematical Logic, to which Dana Scott was appointed. The professorship and readership in mathematical logic had been established within the Philosophy Sub-Faculty, but in 2000 were transferred to the Mathematical Institute, where they now constitute the core of a world renowned group in model theory. A significant number of the best graduates in philosophy from Oxford have come from the honour school of Mathematics and Philosophy since its founding.

In 1968 Dummett was elected a Fellow of the British Academy.[32] He had at that stage adumbrated a philosophical programme in eight published papers, which he would then pursue over the next forty-three years. This election came at the end of a period of four years in which he had given up writing philosophy altogether in order to devote himself to the fight against racism. Years later he wrote, 'I thought at the time that I had wrecked my career, as did Ann, but I was content that the sacrifice was

---

[31] Ibid., p.17.

[32] At forty-three Dummett was the second youngest in the cohort of twenty-one new Fellows elected in that year, whose average age was fifty-six, but Timothy Smiley, writing in 1995, noted, in regard to the election of William Kneale in 1950 at the age of forty-four, that 'to its credit the philosophy section of the Academy recruits new members a decade younger than their opposite numbers in other subjects' (T. J. Smiley, 'William Calvert Kneale, 1906–1990', *Proceedings of the British Academy*, 87 (1995), p. 386). In recent years the ages at which philosophers have been elected to the Academy have tended to be as in other subjects.

worth it, the enemy being so evil. Some years later I discovered that I had not after all wrecked my career.'[33] Election to the British Academy indeed signalled that he had been able to devote himself to anti-racism for the preceding four years while remaining a potent force in the development of philosophy.

In 1984 he resigned his Fellowship of the British Academy in protest at the Academy's failure, as he saw it, to stand up to the Thatcher government's attack on British universities by its cuts to spending for higher education and research, as he explained in a letter published in *The Guardian* on 19 June 1984, under the headline 'When an academy leaves academics in the lurch.' He declared that

> The universities have very few friends among politicians, journalists, or any other external group. Their champions ought to be the academies that exist to foster research in academic disciplines which—especially in the arts—is very largely carried out in the universities. Of these, the British Academy covers all academic disciplines other than the sciences; in the face of the Government's unprecedented attack upon the universities, it has been its evident duty to defend their cause. […] It has made no adequate attempt to fulfil this duty.

Dummett absolved the Fellows of the Academy of responsibility for this situation, and laid blame on those running the Academy: 'The Fellows as a body have little say in what the Academy does, since it is run in a thoroughly undemocratic fashion.'[34] Dummett's public letter had been preceded by a letter of resignation on 2 January 1984, from which he had been temporarily dissuaded, in which he had given as his reason 'the utterly undemocratic nature of the institution', and citing in particular the Blunt affair from four years earlier. (Dummett considered that the then President had gone against the wishes of the Fellows, as expressed in a vote in the 1980 AGM, on a motion put by Dummett, not to ask Anthony Blunt to resign his Fellowship of the Academy in the aftermath of having been identified as a Soviet agent.[35]) Dummett was also disaffected towards the Academy at that time by its refusal to support his research interests in Tarot cards and games. In 1995 Dummett accepted re-election to the British Academy as a Senior Fellow, perhaps in part persuaded, and at any rate not put off, by Timothy Smiley's argument that this would give him the possibility of resigning again over another issue, should one arise.

[33] Dummett, 'Intellectual autobiography', p. 19.
[34] Ibid.
[35] For another view of this affair see the section by P. W. H. Brown in D. A. Russell and F. S. Halliwell, 'Kenneth James Dover 1920–2010', *Biographical Memoirs of Fellows of the British Academy*, 11 (London, 2012), pp. 169–71.

Later career

The period in which Dummett gave the fight against racism highest priority among all his commitments came to an end in 1968. As he explained

> by 1968, Britain had become irretrievably identified by the black people living here as a racist society …. The alienation of racial minorities is now so great that a white ally in the struggle can, except in special circumstances, play only the most minor ancillary part. It was only at the stage at which … I felt that I no longer had any very significant contribution to make, that I thought myself justified in returning to writing about more abstract matters of much less importance to anyone's happiness or future.

Dummett offered this account of his return to writing philosophy in the Preface to his first book, *Frege: Philosophy of Language*,[36] published in 1973, to great critical acclaim. Dummett went on to publish eight further books in philosophy and three volumes of essays In the same year as he published his first book, his wife Ann published *A Portrait of English Racism*,[37] about which Dummett later said, 'I would rather have written that book than any of the many I have written.'[38]

In 1974 Dummett applied for and was elected to a Senior Research Fellowship at All Souls, and resigned as Reader in the Philosophy of Mathematics, in order to have more time for research and to be free to work more broadly than specifically in the philosophy of mathematics. In this period he had embarked on a series of major philosophical papers pursuing lines of research adumbrated in 'Truth', beginning in 1973 with his British Academy lecture 'The justification of deduction', and continuing with 'The philosophical basis of intuitionistic logic' (given as a lecture in 1973, published in 1975), 'What is a theory of meaning?' (lecture in 1974, published in 1975), 'What is a theory of meaning? (II)', and his William James Lectures at Harvard, 'The logical basis of metaphysics' (given in 1976, published in an expanded and revised form as a book in 1991).[39]

In 1977 Dummett published *Elements of Intuitionism*,[40] a remarkable accomplishment mathematically, philosophically, and pedagogically. He there established that

[36] M. A. E. Dummett, *Frege: Philosophy of Language* (London, 1973), pp. x–xi.
[37] A. Dummett, *A Portrait of English Racism* (Harmondsworth, 1973).
[38] Dummett, 'Intellectual autobiography', p. 24.
[39] M. A. E. Dummett, 'The justification of deduction', *Proceedings of the British Academy*, 59 (1974), 201–32; M. A. E. Dummett, 'The philosophical basis of intuitionistic logic', in H. E. Rose and J. C. Shepherdson (eds.), *Logic Colloquium '73: Proceedings of the Logic Colloquium Bristol, July 1973* (Amsterdam, 1975), pp. 5–40; M. A. E. Dummett, 'What is a theory of meaning', in S. Guttenplan (ed.), *Mind and Language* (Oxford, 1975), pp. 97–138; M. A. E. Dummett, 'What is a theory of meaning (II)', in G. Evans and J. McDowell (eds.), *Truth and Meaning: Essays in Semantics* (Oxford, 1975), pp. 67–137; and M. A. E. Dummett, *The Logical Basis of Metaphysics* (Cambridge, MA, 1991).
[40] M. A. E. Dummett, *Elements of Intuitionism* (Oxford, 1977), (2nd edn, Oxford, 2000).

intuitionist mathematics and logic can indeed be cast in the form of Dummettian anti-realism, as foreshadowed in 'Truth'—a completely different basis from the psychologism by which Brouwer had argued for intuitionist mathematics.

In 1979 Dummett was elected to the Wykeham Professorship of Logic, and moved from All Souls, which had been his academic home for twenty-nine years, to New College, with which the Wykeham chair is associated. The question in that election was not whether Dummett would be offered the job but whether he would accept, which entailed giving up his Research Fellowship at All Souls, with its very limited formal demands, which he had held for five years and which could have continued for thirteen years more, until retirement. His taking up the chair was a selfless act of loyalty to Oxford Philosophy. Almost immediately he was called upon to supervise substantially more than fifteen graduate students at a time. This was in part because professors had a statutory obligation to do a lot of graduate supervision, but mostly because his publications were now setting the agenda for important philosophical developments, and graduates flocked to Oxford to study with him.

In 1982 Dummett was awarded a Humboldt-Stiftung Research Prize which he used for four months at the University of Münster, working on Frege. He spent the academic year 1988–89 in Stanford as a Fellow of the Center for Advanced Study in the Behavioral Sciences, during which he finished two major books begun earlier, *Frege: Philosophy of Mathematics* and *The Logical Basis of Metaphysics* (both published in 1991).[41] In 1989 Dummett took the Oxford degree of DLitt (a higher doctorate awarded on the basis of publications).[42]

Final years

Michael Dummett retired from Oxford in 1992, at the compulsory age of sixty-seven. He gave many lectures in retirement, including the Gifford Lectures at St Andrews University in 1997, which he published as *Thought and Reality* in 2006.[43] His aim in those lectures and the ensuing book was 'to describe the conception of the world—of reality—that would be proper to one who accepted the version of anti-realism that has been associated with me, namely a generalisation to all language of the intuitionist understanding of mathematical language, which I have never for long more than provisionally accepted. It turned out very Berkeleian, with a strong asymmetry between past and future, something to which I am temperamentally averse.'[44] In 2002

[41] M. A. E. Dummett, *Frege: Philosophy of Mathematics* (London, 1991); Dummett, *The Logical Basis of Metaphysics.*
[42] 'Dummett, Sir Michael (Anthony Eardley)', *Who's Who.*
[43] M. A. E. Dummett, *Thought and Reality* (Oxford, 2006).
[44] Dummett, 'Intellectual autobiography', p. 31.

he gave the John Dewey Lectures at Columbia University, published in 2004 as *Truth and the Past*,[45] in which he continued the struggle with which he had been engaged ever since his paper 'Truth' between the pull toward a global anti-realism and a countervailing pull toward a realist view on statements about the past, as he had explored in 'The reality of the past', in 1969, where his final sentence had been, 'Of course, like everyone else, I feel a strong undertow towards the realist view: but then, there are certain errors of thought to which the human mind seems naturally prone.'[46] He now attempted again to find a tenable antirealism for statements about the past. His assessment of his earlier attempt in 1969 was that 'the conclusion that I reached was the most disappointing possible. Antirealism about the past was not incoherent; but it was not believable, either. I have been perplexed by this matter ever since.'[47]

Dummett's final philosophical project was to write replies to the twenty-seven essays on his work in the *Library of Living Philosophers* volume on *The Philosophy of Michael Dummett* (which he described as 'sometimes like an experience we are all denied, writing thank-you letters for favourable obituaries'),[48] and to write his 'Intellectual autobiography' for that volume, which he wrote mostly in 2000. His replies get to the heart of the various matters under discussion and constitute an invaluable resource for understanding his thinking. The volume appeared in the summer of 2007; Dummett said then that he no longer felt able to do any new philosophy, though he continued to attend the philosophy of mathematics seminar in Oxford until the spring of 2010. He died on 27 December 2011, at the age of eighty-six, four days before what would have been his and Ann's 60th wedding anniversary. He was buried in Wolvercote Cemetery on 17 January 2012 after a Requiem Mass at St Aloysius Church. Ann died six weeks after Michael, on 7 February 2012, and they were commemorated together in a memorial service in New College Chapel on 2 June 2012.

Honours

Among honours not already mentioned, Dummett received five honorary degrees (University of Nijmegen 1983; University of Caen 1993; University of Aberdeen 1993; University of Stirling 2002; University of Athens 2004). He was elected Honorary Foreign Member of the American Academy of Arts and Sciences in 1985, and member of the Academia Europaea in 1990. He received the Lakatos Prize in

[45] M. A. E. Dummett, *Truth and the Past* (New York, 2004).
[46] M. A. E. Dummett 'The reality of the past', *Proceedings of the Aristotelian Society*, n.s. 69 (1968–1969), 258.
[47] Dummett, *Truth and the Past*, p. 45.
[48] Dummett, in Auxier and Hahn, *The Philosophy of Michael Dummett*, p. 819.

1994, for his book *Frege: Philosophy of Mathematics*, the Rolf Schock Prize for Logic and Philosophy in 1995, and the Lauener Prize for Analytical Philosophy in 2010. In 2017 Christ Church named its lecture theatre in his honour, thereby bringing the name of Michael Dummett into everyday use in the place where his long and illustrious career in Oxford had begun.

Dummett's character

In 'Truth' Dummett considers the sentence 'Jones was brave or he was not', said of a man who never faced danger in his lifetime, and concludes that 'anyone with a sufficient degree of sophistication' will reject the claim that, on the basis of Jones' character, one or other of these statements is true. Whichever it would be 'must be true in virtue of the sort of fact we have been taught to regard as justifying us in asserting it', not something 'of which we can have no direct knowledge'.[49] Dummett's life was filled with visible manifestations of bravery, combined with great independence of spirit.

Dummett's conversion to Catholicism in the face of disapproval from parents and teachers clearly showed bravery and independence. Doubtless this was so also when he proclaimed his anti-racist views in Malaya among the British colonials there after the war, and Robert Pring-Mill, who served with Dummett in Malaya, recalled him there as displaying 'beatific disregard of danger'.[50] His courage and independence of mind is also shown in the fact that he established himself as a major figure in Oxford philosophy without the support of and indeed in antipathy to its leading figures in his philosophical youth, Gilbert Ryle and John Austin. ('I never greatly cared for Ryle; he tried to make us narrower and narrower, scorning not only Heidegger, whom he had once reviewed respectfully, but Carnap as well',[51] and referring to the fact that he began his philosophical career thinking of himself as a follower of Wittgenstein, 'it helped to inoculate me against the influence of Austin; although he was himself unquestionably a clever man, I always thought that the effect of his work on others was largely harmful, and therefore regretted the nearly absolute domination that for a time he exercised over Oxford philosophy'.[52]) Dummett put himself on the line in the fight against racism, bravely facing great hostility from racists as illustrated in his account of being arrested, charged and tried, with the possibility, averted by acquittal, of being sent to prison, for picketing a hairdresser's in Oxford which refused to serve

---

[49] Dummett, 'Truth', 158–9.
[50] Kenneth Wachter (pers. comm.).
[51] M. A. E. Dummett, 'Reply to Brian McGuinness', in Auxier and Hahn, *The Philosophy of Michael Dummett*, p. 52.
[52] Dummett, *Truth and Other Enigmas*, p. xii.

Asian or Caribbean people;[53] see also *The Times* 8 October 1968, for a photo of Dummett confronted by a steward when protesting during the meeting of the 'Society for Individual Freedom'. Michael and Ann received death threats during this time, in response to which, on the advice of the police, sheets of bullet-proof clear plastic were put over the windows at the back of their house. None of these dangers deflected Dummett from his pursuit of anti-racism.

Moral outrage at flagrant injustice or culpable irresponsibility or cruelty could rouse him to fury, seldom manifested, but volcanic in its occurrence, like Vesuvius, which has long periods of quiet followed by an almighty explosion, rather than Etna, which is in a state of more or less continuous eruption, as Timothy Smiley noted. (Dummett's resignation from the Academy may be seen as one of those Vesuvian eruptions.) Far more characteristic of Dummett than his fury was his kindness and compassion, and great generosity. His brilliance as a philosopher was matched by brilliance as a teacher, and the generosity of his commitment to anti-racism was matched by generosity towards his students and colleagues, with his time and his ideas, and his engagement with their ideas. When he died, he was warmly remembered in an unprecedented collective expression of gratitude, affection, and admiration by twenty-six members of the philosophy profession, many of them his former students (including the two authors of this memoir) in the Opinion Pages of the *New York Times* on 4 January 2012.[54] At the centre of Dummett's life was his family, and many affectionate memories of his students and friends are of being generously welcomed by Michael and Ann into their family home. Dummett's jovial good humour and infectious laugh, combined with the depth and humanity of his conversation, enriched the lives of those around him.

Philosophical work

Reflecting on the nature of the progress in philosophy, Dummett suggests that 'the path toward the goal of philosophy—any path that we may take—is a meandering one that twists and turns upon itself. At a given stage, the only way to proceed any further along this path may be to go quite a long way in a direction opposite to that in which the goal lies; to go in that direction may be the only way to improve our chance of eventually reaching the goal.'[55] Dummett was true to this precept in being unusually

---

[53] Dummett, 'Intellectual autobiography', p. 20.
[54] https://opinionator.blogs.nytimes.com/2012/01/04/remembering-michael-dummett/ (accessed 13 September 2018).
[55] M. A. E. Dummett, *The Nature and Future of Philosophy* (New York, 2010), p. 149 (first published in Italian translation, 2001).

unconcerned with whether interlocutors agreed with him about the ultimate answers to philosophical questions. Those answers, he thought, were no more than highly fallible predictions about where the path might eventually lead. What mattered was using one's machete to clear away the tendrils of confusion that obscure any philosophical issue and then taking some further steps down the path. The assessment of Dummett's philosophical work which follows has been written in the same spirit. We have not been afraid to say which trails seem to us to lead to dead ends. Only by doing so will philosophers be encouraged to concentrate their efforts at the points where further progress is likely.

Early papers

Dummett began to publish in 1954 and over the next ten years published thirteen full-length papers (six of them in *The Philosophical Review*) alongside four substantial reviews. Between 1965 and 1973, by contrast, he published only a short encyclopaedia article on Frege and the paper 'The reality of the past', which was read to the Aristotelian Society in 1969. Accordingly, the articles which appeared between 1954 and 1964 constitute a distinctive part of his oeuvre, one on which his early international reputation rested.

In style, these papers are typical of their time and differ from Dummett's later publications. They are short and largely free of footnotes: readers were trusted to know the literature which an author might have in mind. The writing is precise but terse; the reader is also expected to fill in some vital argumentative steps.[56] There is throughout a strong sense of a powerful, fresh, and logically ingenious mind addressing itself to a wide range of philosophical topics.

The philosophy of time was an early and enduring preoccupation: along with the possibility of retro-causation, Dummett defended McTaggart's notorious argument that time was unreal. Traditional metaphysical concerns also loom large: a critical notice and two essays reflect an intense early engagement with Nelson Goodman's *The Structure of Appearance*.[57] Although he admired the book's technical adroitness, Dummett had little sympathy with Goodman's neo-Carnapian project of constructing the world from qualia. One of the essays on Goodman, 'Nominalism', also shows the importance that Frege's Context Principle ('Only in the context of a sentence does a word have meaning'[58]) had already come to assume for Dummett. Dummett understands the Principle to imply that 'if a word functions as a proper name, then it *is* a

---

[56] Dummett may on occasion have expected too much. The Appendix, below, draws on personal communication with him to explicate an argument that is merely sketched in his 1964 paper, M. A. E. Dummett, 'Bringing about the past', *The Philosophical Review*, 73 (1964), 338–59.

[57] N. Goodman, *The Structure of Appearance* (Cambridge, MA, 1951).

[58] M. A. E. Dummett, 'Nominalism', *The Philosophical Review*, 65 (1956), 491.

proper name'.[59] He takes this to exclude the sort of nominalism espoused by Goodman and (at one stage) by Quine, which allows that the numeral '28' functions as a proper name, and that the statement '28 is a perfect number' is true, but nonetheless there is no such thing as the number 28.

The early papers also include two contributions to formal logic. 'A propositional calculus with denumerable matrix'[60] explores a logic, *LC*, in which the schema $(A \rightarrow B) \vee (B \rightarrow A)$ is added to the axioms of the intuitionistic propositional calculus. The main result is that *LC* (now called Gödel-Dummett logic[61]) is complete with respect to any infinite lattice with zero and unit elements whose constitutive partial ordering is linear. 'Modal logics between S4 and S5' (1959), written in collaboration with E. J. Lemmon,[62] showed that the modal system S4.2, got by adding the axiom schema $\Diamond \Box A \rightarrow \Box \Diamond A$ to the familiar system S4, has five distinct affirmative modalities (i.e.. $\Box A, \Diamond \Box A, \Box \Diamond A, A, \Diamond A$). It also introduced the fruitful notion of an 'order closure algebra', which Dummett later renamed a '*QO*-space'. *QO*-spaces are close relations of the 'frames' which Saul Kripke used in giving his celebrated semantic theories for modal and intuitionistic logics.

In this period, Dummett also began to make his mark in the philosophy of mathematics. His 1959 assessment of Wittgenstein's contribution to that field brought to the fore Wittgenstein's discussions of following a rule. Dummett, however, became dissatisfied with his understanding of Wittgenstein and wrote, towards the end of his life, 'I should like to come to terms with Wittgenstein: I am sure I have not yet.'[63] In 'The philosophical significance of Gödel's theorem', he developed the difficult but suggestive concept of indefinite extensibility (see below p. 221).[64]

'Truth' and the anti-realist programme

Important as many of these pieces were, they are overshadowed by 'Truth'. This paper is Dummett's first published attempt to address the questions about meaning, logic and realism that were to dominate his philosophical thinking to the end of his life. Some of its suggestions did not bear fruit in Dummett's later writings; he did not

---

[59] Ibid., p. 494.

[60] M. A. E. Dummett, 'A propositional calculus with denumerable matrix', *The Journal of Symbolic Logic*, 24 (1959), 97–106.

[61] R. Dyckhoff, 'A deterministic terminating sequent calculus for Gödel-Dummett logic', *Logic Journal of the IGPL*, 7 (1999), 319–26.

[62] M. A. E. Dummett and E. J. Lemmon, 'Modal logics between S4 and S5', *Mathematical Logic Quarterly*, 5 (1959), 14–24.

[63] Dummett, 'Reply to Brian McGuinness', p. 54.

[64] M. A. E. Dummett, 'The philosophical significance of Gödel's theorem', *Ratio* 5 (1963), 140–55.

pursue, for example, the falsificationist theory of meaning adumbrated at pages 149–50 (see also remark (5) in the Postscript to 'Truth' that Dummett wrote in 1972, and the second paragraph of Dummett's 'Reply to Ian Rumfitt'[65]). The paper culminates in the first published statements of two theses to which Dummett remained strongly attracted throughout the rest of his life, even though he was well aware of the problems which confront them. First, the meaning or sense of a statement should not, in general, be given by specifying the conditions under which it is true; rather, it should be given by saying 'when it may be asserted in terms of the conditions under which its constituents may be asserted'.[66] This claim—which Dummett later labelled 'justificationism'—was the basis of a strong form of 'anti-realism' whereby 'the concept of truth-values determined by reality independently of us should be abandoned. The notion of a statement's being true should be replaced by that of its being shown to be true.'[67] The second thesis was that adopting this anti-realist position, whereby truth is 'dethroned' from its central place in the theory of meaning, would in turn require dethroning certain principles of classical logic—notably the Law of Excluded Middle—from their status as logical laws.

In the mid-1970s Dummett set out substantive arguments for these theses. The argument for anti-realism was presented in his two essays entitled 'What is a theory of meaning?', the first of which came out in 1975 with the second appearing the following year. He began to elaborate his case against classical logic—and his argument that intuitionistic logic is the strongest system that can be philosophically justified—in two lectures which were delivered in 1973 and published 1975: 'The philosophical basis of intuitionistic logic' and 'The justification of deduction'. Both the main theses of 'Truth' were also defended in the William James Lectures which Dummett gave at Harvard in early 1976, and published in considerably revised form as *The Logical Basis of Metaphysics* in 1991.

Dummett's main argument for justificationism was that it is the only theory of meaning that makes possible a non-circular account of what it is to understand a statement. The salient contrast is with the more familiar truth-conditional theory, whereby a statement's content is constituted by the conditions under which it is true. According to Dummett, a theory of meaning is of interest only if it is a theory of our knowledge of meaning: as he often put it, 'a theory of meaning is a theory of understanding'.[68] On this conception, the key thesis of the truth-conditional theory is the

[65] Dummett, *Truth and Other Enigmas*, p. 22; M. A. E. Dummett, 'Reply to Ian Rumfitt', in Auxier and Hahn, *The Philosophy of Michael Dummett*, p. 694.
[66] Dummett, 'Truth', 161.
[67] Dummett, 'Intellectual autobiography', p. 18.
[68] Dummett, 'What is a theory of meaning?', p. 99; this is reprinted in M. A. E. Dummett, *The Seas of Language* (Oxford, 1993), pp. 1–33 at p.3.

claim that understanding a statement is a matter of knowing under what conditions it is true. In general, however, this knowledge will be implicit, and Dummett held that an ascription of implicit knowledge to a speaker is vacuous unless it amounts to attributing to him a disposition, the possession of which may be fully manifest in his behaviour. However, knowledge of the conditions under which a statement is true cannot always be cashed out as a fully manifest disposition. At least, this is so if our conception of truth is the usual realist one, whereby a statement may be true in circumstances where no one can recognise that it is true. For, Dummett supposed, the only plausible candidate to be a disposition, possession of which amounts to knowing a statement's truth conditions, is the disposition to assent to it in circumstances where one recognises that it is true. And if a statement may be true in circumstances where no one can recognise that it is true, then a speaker's implicit knowledge that a state-ment is true in, and only in, certain conditions will not be fully manifest in his dispo-sition to assent to it in the circumstances in which it may be recognised as true. By contrast, Dummett claimed, knowledge of the conditions in which we have grounds for asserting the statement may be fully manifest in a speaker's behaviour: such know-ledge will be manifest in the speaker's disposition to assert the statement when he has such grounds, and to refrain from asserting it when he does not.

Critics challenged this argument at a number of points. In the eyes of many, the main premiss—that a speaker's knowledge of meaning must be fully manifest in his behaviour—was more a philosopher's prejudice about how language ought to work rather than anything that can be applied in analysing actual linguistic practice, though in 'Truth' Dummett says of his doctrine that 'we no longer explain the sense of a state-ment by stipulating its truth-value in terms of the truth-values of its constituents, but by stipulating when it may be asserted in terms of the conditions under which its constituents may be asserted. The justification for this change is that this is how we in fact learn to use these statements.'[69] Another worry is that this requirement of full manifestability is so strong that even a justificationist cannot meet it. We will consider briefly Dummett's attempts to allay the first doubt by constructing justificationist semantic theories for reasonably large fragments of a natural language.

As the quotation from 'Truth' in the previous paragraph already shows, Dummett differed from the verificationists of the Vienna Circle in taking seriously the composi-tionality of linguistic content. Like most empirical linguists, he held that the content of a complex statement is determined by the contents of its parts. Accordingly, he had to show (at least in outline) how a compositional justificationist semantic theory would go. His model here was the semantic theory for the language of intuitionistic mathematics that had been proposed by Arend Heyting. As Dummett explains this theory,

[69] Dummett, 'Truth', 161.

the meaning of each [logical] constant is to be given by specifying, for any sentence in which that constant is the main operator, what is to count as a proof of that sentence, it being assumed that we already know what is to count as a proof of any of the constituents. The explanation of each constant must be faithful to the principle that, for any construction that is presented to us, we shall always be able to recognize effectively whether or not it is a proof of any given statement. For simplicity of exposition, we shall assume that we are dealing with arithmetical statements…

The logical constants fall into two groups. First are $\wedge$, $\vee$ and $\exists$. A proof of $A \wedge B$ is anything that is a proof of A and of B. A proof of $A \vee B$ is anything that is a proof either of A or of B. A proof of $\exists x A(x)$ is anything that is a proof, for some n, of the statement $A(\bar{n})$.

The second group is composed of $\forall$, $\rightarrow$, and $\neg$. A proof of $\forall x A(x)$ is a construction of which we can recognize that, when applied to any number n, it yields a proof of $A(\bar{n})$. Such a proof is therefore an *operation* that carries natural numbers into proofs. A proof of $A \rightarrow B$ is a construction of which we can recognize that, applied to any proof of A, it yields a proof of B. Such a proof is therefore an operation carrying proofs into proofs…A proof of $\neg A$ is usually characterized as a construction of which we can recognize that, applied to any proof of A, it will yield a proof of a contradiction.[70]

This semantic theory explains why certain classical logical laws are not logically valid for the intuitionist. A statement will count as intuitionistically valid if the semantic principles guarantee it to be provable no matter which atomic statements are provable. So a statement in the form $\ulcorner A \vee \neg A \urcorner$ will be valid only if either A or its negation is provable. Since it cannot be assumed of an arbitrary statement that either it or its negation is provable, Excluded Middle is not an intuitionistic logical law.

Heyting's semantics, though, needs to be generalised before it can be applied to a natural language, few of whose statements admit of anything that can properly be called a 'proof'. Since a mathematical proof justifies its conclusion, Dummett opted for a generalisation in which the semantic axiom for 'or' is as follows (and similarly for the other connectives):

(J) A justification of $\ulcorner A$ or $B \urcorner$ is anything that is a justification either of A or of B.

As Dummett acknowledged, (J) is untenable if it is understood to concern an individual's justification for his assertions, at a particular time: 'I may be entitled to assert $\ulcorner A$ or $B \urcorner$ because I was reliably so informed by someone in a position to know, but if he did not choose to tell me which alternative held good, *I* could not [assert either disjunct].'[71] For this reason, he understood (J) as concerning the existence of justifications, not a

[70] Dummett, *Elements of Intuitionism*, p. 12 (2nd edn., p. 8).
[71] Dummett, *The Logical Basis of Metaphysics*, p. 266.

given thinker's apprehension of them. All the same, if a justification is to exist, he required that a suitably placed thinker *could have* apprehended it, even if none in fact did.[72]

Even when understood in this way, however, another of Dummett's own examples points to a whole range of cases where (*J*) remains problematic. 'Hardy may simply not have been able to hear whether Nelson said, "Kismet, Hardy" or "Kiss me, Hardy", though he heard him say one or the other: once we have the concept of disjunction, our perceptions themselves may assume an irremediably disjunctive form'.[73] On its face, this is a counterexample to (*J*). If Hardy heard Nelson say one thing or the other, then there was—indeed, he had—very strong justification for asserting the disjunction. But in the circumstances of the Battle of Trafalgar, there may have been no justification that any observer could have apprehended for asserting either disjunct. Hardy was as well placed to hear Nelson's last words as anyone could have been, but all he could hear was that Nelson said either one thing or the other. Dummett's last point is the crucial one: our perceptions may themselves assume an irremediably disjunctive form. Since they may do so, the only way to protect (*J*) from this sort of counterexample is to deny that perceptions can constitute justifications for disjunctive assertions. But in that case (*J*) forces so radical a departure from our ordinary understanding that the notion it characterises is unrecognisable as our notion of disjunction.

In *The Logical Basis of Metaphysics*, Dummett tried to get around this problem by following Dag Prawitz in distinguishing between a statement's *canonical* or *direct* grounds and those which are merely indirect: it is only the statement's direct grounds which give its content. Dummett postulated that a speaker has direct grounds for asserting ⌜*A* or *B*⌝ when, and only when, he has either direct grounds for asserting *A* or direct grounds for asserting *B*. The direct grounds for other complex statements follow this pattern: they embody the standard introduction rule for the statement's principal connective.

This move creates a problem about the validity of deductive proofs. Many proofs (or apparent proofs) of complex statements do not terminate with an application of the introduction rule for the conclusion's principal connective. Given that a statement's content is given by its canonical grounds, it seems that such proofs (or apparent proofs) will be unfaithful to the contents of their conclusions. Dummett maintained that they may be faithful so long as they show how to transform any direct grounds for all the premises into a direct ground for the conclusion. Indeed, he took this condition to be the criterion for an argument to be deductively valid. In this way, truth

[72] Ibid., p. 268.
[73] Ibid., p. 267.

was 'dethroned' not only from its central place in the theory of meaning, but also from its traditional place in the explanation of validity.

Dummett illustrated this anti-realist conception of validity by using Euler's famous solution to the problem about the bridges of Königsberg. Euler's proof is valid in that it 'show[s] us, of someone observed to cross every bridge at Königsberg, that he crossed at least one bridge twice, *by the criteria we already possessed for crossing a bridge twice*'.[74] 'When an expression, including a logical constant, is introduced into the language, the rules for its use should determine its meaning, but its introduction should not be allowed to affect the meanings of sentences already in the language.'[75] By mastering logical rules, we acquire new indirect grounds for making assertions, even of atomic statements. However, the conditions in which atomic statements may be directly asserted, and hence their meanings, are not disturbed.

This account of validity generates serious problems of its own. Euler's proof is said to show, of someone observed to cross every bridge at Königsberg, that he crossed at least one bridge twice, *by the criteria we already possessed for crossing a bridge twice*. But that cannot mean that those criteria were actually applied to verify that the promenader crossed a bridge twice. Perhaps they were—perhaps an observer stationed on the Dombrücke, for example, saw the promenader cross that bridge twice—but the proof would not be invalidated if the pre-existing criteria were not actually applied. The most that can be claimed is that the proof's validity consists in the truth of a counterfactual claim: had an observer been stationed on each bridge, with instructions to tick a box if, and only if, the promenader was observed crossing it twice, at least one observer would have ticked his box.

This analysis, however, is susceptible to objections parallel to those which face putative counterfactual analyses of other notions. There are possible worlds in which all the inhabitants of Königsberg are afflicted by Königsberg ennui, a neurological condition which ensures that anyone trying to observe whether a promenader has crossed a given bridge twice falls into a catatonic state before any second crossing. In such a world, it will not be true to say that at least one of the observers would have ticked his box, had the promenader crossed every bridge at least once. Even in such a world, though, Euler's deduction is valid.

Other doubts about the form of anti-realism proposed in *The Logical Basis of Metaphysics* arise when we reflect on the role which the distinction between direct and indirect grounds needs to play in it. The notion of directness needs to be sufficiently generous that no ground for asserting a formula obtains unless a direct ground for asserting it could have obtained. Yet the direct grounds for asserting a complex

---

[74] Ibid., p. 219, emphasis in the original.
[75] Ibid., p. 220.

formula are constrained to be those given by the introduction rule for the formula's main connective. Combining these points, we deduce that no ground for asserting a complex formula can obtain unless the assertion of that formula could have been justified by applying the introduction rule for its main connective. This thesis is what Dummett calls his *Fundamental Assumption* and it opens the way to a new assault on classical logic, one which does not require accepting Heyting's semantic clauses for the connectives and quantifiers. The Assumption combines with the account of deductive validity to yield the requirement that the introduction and elimination rules for a given connective must be 'in harmony'. Dummett contended that, while the intuitionistic rules for negation possess this virtue, the classical rules do not.

The basic problem with this line of attack on classical logic is that the Fundamental Assumption is highly doubtful. Dummett concluded his own discussion of it by saying that 'our examination of the fundamental assumption has left it very shaky'.[76] While it may be tenable for the case of conjunction, it is indeed implausible for all the other sentential connectives, and particularly for the key case of negation. According to the Assumption, we shall not be entitled to assert a negated statement unless we could have justified it by applying the introduction rule for 'not'. That rule licenses the assertion of ⌜Not A⌝ when a contradiction has been derived from our premises along with the hypothesis A. In many circumstances where we take ourselves to be entitled to assert ⌜Not A⌝, though, it is hard to see what the appropriate premises might be. Suppose you look out of the window and see that it is not raining. You are surely entitled to assert 'It's not raining', but in many circumstances your observation delivers no premises that would enable you to justify your assertion by applying the rule of 'not'-introduction. In looking out of the window, you might see that it is sunny, but being sunny is compatible with rain. The only specification of the content of your experience that is guaranteed to be incompatible with 'It is raining' is 'It is not raining', but while you can indeed see that it is not raining, the belief that it is not raining serves as a *premiss* in your reasoning. It is not a conclusion which has been reached by applying the rule of 'not'-introduction.

For the reasons set out, Heyting's semantic theory does not seem to generalise so as to yield a plausible account of the meanings of empirical statements. It does not follow at all, however, that justificationism is doomed; there are justificationist semantic theories which do not take Heyting's semantics as their model.

Whether any such theory is really viable, and whether it can sustain classical logic, remain important open questions in the philosophy of language. One class of statements which present a particular challenge to the coherence of global anti-realism are those about the past. Dummett early recognised the problem they pose. In his

---

[76] Ibid., p. 277.

important paper 'The reality of the past', he wrote: 'I think that without doubt the thorniest problem for one who wishes to transfer something resembling the intuition-ist account of the meanings of mathematical statements to the whole of discourse is what account he can give of the meanings of tensed statements.'[77] The problem arises from the apparent 'existence of the truth-value link',[78] that what is true at a certain time remains true, regardless of whether or not the evidence that showed it to be true at that time is later irretrievably lost. 'No matter what manoeuvres he attempts, the anti-realist will be unable to avoid inconsistency in recognising the existence of the truth-value link if he formulates his contention as being that a past-tense statement, made at any given time, is true at that time only if there is at that time a situation jus-tifying the assertion of the statement.'[79] How uncertain he felt about the position he was attempting to maintain comes out poignantly in the last line of the paper: 'Of course, like everyone else, I feel a strong undertow towards the realist view: but, then, there are certain errors of thought to which the human mind seems naturally prone.'[80] Dummett returned to this problem time and again, and never reached a settled view on how to solve it. He adumbrated incompatible solutions in his final two books, *Truth and the Past* and *Thought and Reality*. These two books started life as invited lecture series: *Thought and Reality* began as the Gifford Lectures at St Andrews, in 1996; *Truth and the Past* began as the Dewey Lectures at Columbia, given in 2002. (One further book by Dummett appeared in his lifetime, *The Nature and Future of Philosophy*, published in 2010, but he had already written it by 2001, when it was published in Italian translation.)

Dummett was perfectly clear that his proposed solutions to this problem in these two sets of lectures, and their subsequent publications (in the reverse order from their delivery) were incompatible, as he spells out in the Preface to *Thought and Reality*:

> In the Gifford Lectures, a proposition is reckoned to be true just in case *we* [Dummett's emphasis], as we are or were, are or were in a position to establish it to hold good; my present standpoint, as stated in the Dewey Lectures, is that it is true just in case *anyone suitably placed in time and space* would be or have been [the use of the subjunctive here is hugely significant] in such a position. The difference has an evidently far-reaching effect; far more propositions will be rendered true under the Dewey than under the Gifford conception.[81]

[77] Dummett, 'The reality of the post', 250–1.
[78] Ibid., p. 245.
[79] Ibid., p. 251.
[80] Ibid., p. 258.
[81] Dummett, *Thought and Reality*, pp. vii–viii.

Dummett was adamant that such divergences of views in the corpus of a single phil-osopher are in the nature of doing philosophy, as he makes clear in the Preface to *Truth and the Past*:

> The position I have adopted in this book is greatly at variance with those I expressed in my not yet published Gifford Lectures of a few years ago. In those, I did not embrace antirealism about the past: but I did maintain that the body of true statements is cumulative. I have not published these lectures, which it is the normal practice to do, because I was troubled that this view was in error. Now that I am publishing a book expressing a different view, I think I will probably publish the Gifford Lectures as I gave them. I do not think that anyone should interpret everything that a philosopher writes as if it were just one chapter in a book he is writing throughout his life. On the contrary, for me every article and essay is a separate attempt to arrive at the truth, to be judged on its own.[82]

Important books

While the project of combining anti-realism with logical revisionism was Dummett's most distinctive and original contribution to philosophy, he also did significant work in other areas.

Chief among these is Frege scholarship. The first book Dummett published was *Frege: Philosophy of Language*, which appeared in 1973. It offered the earliest system-atic presentation of Frege's doctrines outside the philosophy of mathematics and was widely hailed as a masterpiece. Its interpretation of Frege has been challenged. In particular, some scholars have denied that Frege was, centrally, a philosopher of lan-guage or even that he had a philosophy of language. Dummett's account certainly downplays the extent to which Frege was motivated by epistemological concerns—in particular, by the desire to get clear about what ultimately justifies our acceptance of the basic principles of arithmetic and geometry. The value of the book, though, does not depend on its offering a fully convincing interpretation of Frege's writings. It lies, rather, in Dummett's having created an intellectual framework in which certain key Fregean theses have a secure and comprehensible place, and which enabled him to compare them fruitfully with central contentions in the philosophy of language of the middle fifty years of the twentieth century. Thus the book contains illuminating discussions of, inter alia: Kripke's theory of names as rigid designators; Russell's and Strawson's accounts of definite descriptions; the nature of the difference between particulars and universals; Quine's analysis of belief ascription, of ontological com-mitment, and attack on the analytic/synthetic distinction; Wittgenstein's remarks on

---

[82] Dummett, *Truth and the Past*, p. x.

names, truth, and the speech act of assertion; Prior's analysis of tenses; Moore's account of 'exists'; the nature of abstract objects and the temptations of nominalism; Poincaré's strictures on impredicative definition; and Geach's theory of relative identity. In places, the discussion is prescient. For example, the 'new relativism' that has come to the fore in the past twenty years is both anticipated and criticised.[83] As Dummett realised, shorter and more clearly articulated chapters would have made the book more accessible (a lesson he took to heart in later books), although the provision of a proper index in the second edition (where the first had only a 'Brief Subject Index' and an 'Index of Names') helps. Despite these flaws, the book remains a stimulating source of ideas forty-five years on.

Dummett's second book, *Elements of Intuitionism*, appeared in 1977 (with a second edition in 2000). It includes a pedagogically useful chapter expounding intuitionistic arithmetic and analysis, but the focus is on intuitionistic logic, and the two central chapters, on the formalisation of the logic and on its semantics, interweave formal exposition and philosophical discussion in a most satisfying way. In addition to the originality and clarity of his exposition of intuitionist mathematics and logic in that book, Dummett also established intuitionistically the completeness of negation-free intuitionist logic, a best possible result in light of the result by Gödel and Kreisel that the completeness of Heyting's predicate calculus intuitionistically implies Markov's Principle, which is not intuitionistically valid. This result was obtained independently around the same time, by Harvey Friedman, by different means.[84] In the semantic analysis of intuitionistic logic, Dummett makes effective use of his and Lemmon's notion of a *QO*-space (see p. 209 above) to illuminate the relationship between Heyting's semantic theory and that of E. W. Beth. Dummett remarks in the Preface to the second edition that he has simplified the treatment of valuation systems, which leads to a more perspicuous and elegant presentation of the semantic theory. By contrast, some significant changes of mind on philosophical points are not noted. In particular, the two versions of section 7.2 give very different answers to the vexed question of in what way the intuitionist's account of 'valid proof' must be compositional.

Students of Dummett's philosophy have sometimes been puzzled by his reverence both for Frege, a fervent realist, and for Brouwer, a passionate anti-realist. Comparison of these two books goes a long way to resolve the apparent cognitive dissonance. Brouwer's exposition of intuitionism exemplifies the psychologism which Frege had fiercely attacked. One of Dummett's achievements in *Elements of Intuitionism* was to recast intuitionist mathematics and logic on a completely anti-psychologistic basis. As

---

[83] Dummett, *Frege: Philosophy of Language*, pp. 396–400.
[84] Dummett, *Elements of Intuitionism*, 1st edn. p. 288, 2nd edn. p. 201.

for realism, Dummett remarked that what attracted him to Frege was not his realism, 'for which, I thought, he never really argued, but which he simply took for granted, but the clarity of his thought: much of his thinking was perfectly compatible with a constructive view of mathematics'.[85] Further,

> Reflecting on my rationale for intuitionistic mathematics as an exemplification of Wittgenstein's dictum about meaning as use, it struck me that the metaphysical conceptions accompanying both Platonist and constructive conceptions of mathematics were not the foundations of those conceptions: they were merely pictures illustrating them. One could not argue from the metaphysical pictures, because there was no independent ground for accepting one or the other. The core of the different conceptions lay in the divergent views of what the meanings of mathematical statements must consist in: to adopt one or the other view was to make one or the other picture natural.[86]

Dummett described his third philosophical book as one he never intended to write. *The Interpretation of Frege's Philosophy* appeared alongside the second edition of *Frege: Philosophy of Language* in 1981;[87] it was written to defend the view of Frege advanced in the earlier book against rival conceptions. It evinces a much closer interest than its predecessor in the textual niceties of Frege's writings and, especially, in the context of late nineteenth-century German philosophy. Dummett was there concerned, for example, to downplay the suggestion that Hermann Lotze (a figure who had gone unmentioned in *Frege: Philosophy of Language*) had much influenced Frege.[88] However, the book also contains material of wider philosophical interest. Dummett continues his debate with Kripke over the semantics of proper names, and two chapters pursue a fruitful discussion with Geach. In his critical notice of *Frege: Philosophy of Language*,[89] Geach had objected to the central role Dummett had ascribed to a distinction between simple and complex predicates; on Geach's view, the very distinction, and the related distinction between a statement's 'analysis' and its various 'decompositions', were 'radically unFregean'. In Chapters 15 and 16 of *The Interpretation of Frege's Philosophy*, Dummett convincingly argued that these distinctions, whilst not drawn explicitly, are needed to make best sense of the passages where Frege writes of the 'parts' of thoughts. He also contended forcefully that some such distinction is needed to relate our understanding of statements to their logical powers.

---

[85] Dummett, 'Intellectual autobiography', p. 15.
[86] Ibid., p. 17.
[87] M. A. E. Dummett, *The Interpretation of Frege's Philosophy* (London, 1981).
[88] An irony that amused Dummett was that he had provided the earliest solid evidence that Frege had so much as read Lotze. His 'Frege's Kernsätze zur Logik' (*Inquiry*, 24 (1981), 439–48) shows that the eponymous fragment in Frege's *Nachlass* is a commentary on parts of Lotze's *Logik*.
[89] P. T. Geach 'Critical notice of M. Dummett *Frege Philosophy of Language*', *Mind*, 85 (1976), 436–49.

These chapters show how close attention to what is implicit in a great philosopher's writings can yield insights into the first-order questions he or she was addressing.

Dummett's heavy workload as Wykeham Professor of Logic meant that he published comparatively little during the 1980s. In 1991, however, with a sabbatical at Stanford behind him and retirement imminent, he brought out two new books. As well as *The Logical Basis of Metaphysics* (discussed above), that year saw the appearance (after a twenty-year gestation) of *Frege: Philosophy of Mathematics*. Dummett had originally envisaged this work as one which would compare Frege's theories with the leading twentieth-century approaches to the subject, somewhat as *Frege: Philosophy of Language* had done. He found, however, that Frege's conception of the philosophy of mathematics, which recognised formalism and subjective idealism as the only alternatives to his preferred Platonism, made comparisons with later writers strained and unfruitful. In particular, Frege's view of the field left no room for the sort of intersubjective anti-realism about mathematics that Dummett himself propounded. For this reason, the book that eventually emerged took a very different form. The body of it is a close commentary on Frege's *Die Grundlagen der Arithmetik* and on the sections of his *Grundgesetze* where Frege extends his logicist programme to real analysis. Because it is partly a commentary, *Frege: Philosophy of Mathematics* is the most straightforwardly organised and accessible of Dummett's philosophical books; however, it goes well beyond commentary in tackling problems unsolved by Frege.

The book also entered into controversies. In 1983, Dummett's former student Crispin Wright had published a monograph, *Frege's Conception of Numbers as Objects*,[90] which aimed to revive Frege's logicist project, at least for the arithmetic of the natural numbers, by eschewing the fatal Basic Law V in favour of the abstraction principle specifically concerning cardinal numbers that he had formulated in *Die Grundlagen*. That principle, now called Hume's Principle,[91] says that the cardinal number of *F*s is identical with the cardinal number of *G*s if and only if there is a one-one correlation between the *F*s and the *G*s. This 'neo-Fregean' programme, as it came to be known, owed much to *Frege: Philosophy of Language*. In particular, it drew heavily on Chapter 4 of that book, where Dummett had set out inferential tests for an expression to qualify as a singular term. That debt, though, did not stop Dummett from being sharply critical of the project in *Frege: Philosophy of Mathematics*. Given the formal parallels between Hume's Principle, which is formally consistent, and Basic

[90] C. Wright, *Frege's Conception of Numbers as Objects* (Aberdeen, 1983).
[91] Miscalled, by Dummett's lights: he argued that what Hume had in mind was no abstraction principle, but simply the thesis that two collections of objects are equinumerous if and only there is a one-one correlation between the members of the first collection and those of the second. See the Appendix to M. A. E. Dummett, 'Neo-Fregeans: in bad company?', in M. Schirn (ed.), *The Philosophy of Mathematics Today* (Oxford, 1998), pp. 369–87.

Law V, which is inconsistent in the context of Frege's impredicative second-order logic, the neo-Fregeans were, Dummett thought, too sanguine in assigning to the former principle a foundational epistemological role. He also criticised them for downplaying the importance of the distinction between those singular terms (such as personal proper names), understanding of which is sustained by acquaintance with the bearer, and those (such as numerals) for which such acquaintance is impossible.

Other noteworthy papers

In the period in which Dummett was writing his major books, he also published a number of substantial papers, of which the following three are particularly significant. 'Wang's paradox' (written and presented in 1970, but not published until 1975) anticipates the supervaluationist treatment of vagueness that was to get such attention later in the 1970s. It also identifies some basic flaws in the treatment.[92]

'Is logic empirical?' (1976) analyses Hilary Putnam's famous argument that the paradoxes of quantum mechanics call for revisions to the classical logical law of distribution.[93] Dummett's answer to his titular question is that empirical information could, in principle, bear on logic, but it will do so only via a philosophical account of what logical consequence is. Putnam, Dummett argued, had not justified his preferred account.

'What is mathematics about?' (1993) sketches an anti-realist answer to its titular question.[94] According to Dummett, numbers, sets, and the like are not denizens of a Platonic realm which exists wholly independently of human thinking. Rather, they exist only insofar as thinkers are able to characterise the domains which they compose and share those characterisations with other thinkers. In this third essay, Dummett made extensive use of the notion of indefinite extensiblity he had introduced in 'The philosophical significance of Gödel's theorem' (see above). On his view, some of these fundamental mathematical domains—certainly that of sets—are 'indefinitely extensible'. While this notion has antecedents in the writings of Russell, Poincaré, and Zermelo, it was Dummett's distinctive contribution to contend that quantification over indefinitely extensible domains would not conform to classical logic but would obey only the weaker laws of intuitionistic logic. This 'local' argument for using intuitionistic logic within, for example, set theory is independent of the more general argument against classical logic discussed above. Work by Solomon Feferman and others

---

[92] M. A. E. Dummett, 'Wang's paradox', *Synthese*, 30 (1975), 301–24.
[93] M. A. E. Dummett, 'Is logic empirical?', in H. D. Lewis (ed.), *Contemporary British Philosophy 4th Series* (London, 1976), pp. 45–68.
[94] M. A. E. Dummett, 'What is mathematics about?', in A. George (ed.), *Mathematics and Mind* (Oxford, 1994), pp. 11–26.

on semi-constructive set theory revived interest in this latter argument, which may have a better chance of sustaining its conclusion than the general anti-realist argument.

Dummett's Catholicism and his philosophy

Dummett published twenty articles arising from his deep commitment to Roman Catholicism, of which his most notable contribution to Catholicism and Catholic theology is 'A remarkable consensus', published in *New Blackfriars* in 1987. In this paper, directed against what Thomas Sheehan, Professor of Philosophy at Loyola University in Chicago, had called 'a liberal consensus',[95] Dummett declared that 'the divergence that now obtains between what the Catholic Church purports to believe and what large or important sections of [it] in fact believe ought, in my view, to be tolerated no longer'.[96] This paper generated responses, some virulent, as Dummett experienced them, from five leading Catholic theologians in subsequent issues of *New Blackfriars*, and the debate has continued to be discussed in *New Blackfriars* and in other Catholic publications, for example, *The Annals of Philosophy* of the John Paul II Catholic University of Lublin.[97] It also continued at a conference on Philosophical Theology and Biblical Exegesis held at the University of Notre Dame in March 1990 in which Dummett was invited to give a lecture published as the lead paper in its proceedings as 'The impact of scriptural studies on the content of Catholic belief'.[98]

For most of Dummett's philosophical career his philosophy and his Catholic faith, both of which he pursued passionately, did not appear in the same publications. However, in the *Library of Living Philosophers* volume on *The Philosophy of Michael Dummett*, they are brought together in 'Dummett: philosophy and religion' by Andrew Beards (who at the time of writing was Director of the BA in Philosophy and the Catholic Tradition at Maryvale Institute in Birmingham), which surveys Dummett's writings on Catholicism and Catholic theology and brings them into relation with aspects of his philosophy.[99] Dummett's 'Reply to Andrew Beards', appreciative and mostly in agreement, articulates connections between the two. Particularly striking is a direct connection between the basis of Dummett's anti-realism/justificationism and the necessity to believe in God: 'It makes no sense to speak of a world, or the world,

---

[95] M. A. E. Dummett, 'A remarkable consensus', *New Blackfriars*, 68 (1987), 428.

[96] Ibid., p. 431.

[97] Volume 65 (4), 2017.

[98] M. A. E. Dummett, 'The impact of scriptural studies on the content of Catholic belief', in E. Stump and T. P. Flint (eds.), *Hermes and Athena: Biblical Exegesis and Philosophical Theology* (Notre Dame, IN, 1993), pp. 3–22.

[99] A. Beards, 'Dummett: philosophy and religion', in Auxier and Hahn, *The Philosophy of Michael Dummett*, pp. 863–88.

independently of how it is apprehended. [...] How things are in themselves consists in the way that God apprehends them. That is the only way in which we can make sense of our conviction that there is such a thing as the world as it is in itself, which we apprehend in certain ways and other beings apprehend in other ways.'[100] Dummett also invoked this theism in one of his Gifford Lectures, published as Chapter 8, 'God and the World', in *Thought and Reality*.

Dummett also articulated an understanding of the Catholic doctrine of transubstantiation in which he sought to reconcile 'the fact that no *physical* change occurs at the consecration of the bread and wine'[101] in the Eucharist with the doctrine that after consecration the bread and wine *are* the Body and Blood of Christ, by invoking the notion that after consecration, God *deems* the bread and wine to be the Body and Blood of Christ. Dummett offers an analogy with the adoption of a child: 'When adoptive parents deem an adopted child to be their son, there remains a sense in which he is not their son namely the biological sense', yet 'the parents may legitimately say "He is truly our son."'[102]

Dummett published strong criticisms of the moral teachings of the Catholic Church regarding contraception, which he considered to be gravely mistaken, saying that 'the condemnation of any use of the pill with contraceptive intent by married people is ethically bizarre: an act not wrong in itself is held to be wrong if done for a motive not in itself wrong, indeed often laudable'.[103] He also considered that 'the widely publicized condemnation of the use of condoms in countries where there is a great risk of AIDS is morally objectionable'.[104]

Fergus Kerr concludes his In Memoriam notice for Michael Dummett in *New Blackfriars* with the words, 'Few have ever combined unwavering loyalty to the Church with such relentless interrogation.'[105]

Dummett's place in philosophy

Michael Dummett was one of the most important analytic philosophers of the second half of the twentieth century, among a group that included W. V. Quine, Donald

---

[100] M. A. E. Dummett, 'Reply to Andrew Beards', in Auxier and Hahn, *The Philosophy of Michael Dummett*, p. 892.

[101] Ibid., p. 896

[102] Ibid., pp. 896–7; see also M. A. E. Dummett, 'The intelligibility of Eucharistic doctrine', in W. J. Abraham and S. W. Holzer (eds.), *The Rationality of Religious Belief: Essays in Honour of Basil Mitchell* (Oxford, 1987), pp. 231–61.

[103] Ibid., p. 898.

[104] Ibid.

[105] F. Kerr, 'Michael Dummett In Memoriam', *New Blackfriars*, 93 (2012), 262.

Davidson, P. F. Strawson, Hilary Putnam and Saul Kripke. His espousal of a form of verificationism, distinct in crucial ways from the earlier verificationism of Carnap and the Vienna Circle in the 1920s and first half of the 1930s, and from Quine's later form of empiricist verificationism on the other, started from the question 'In what does grasp of meaning in language consist?'. Few philosophers today espouse Dummett's anti-realist, or justificationist, answer to this question, but his recognition of the central role of that question was and remains enormously influential. The British Academy characterised Dummett's importance in its citation for his re-election to the Academy, as a Senior Fellow, in 1995 as follows:

> Michael Dummett was Wykeham Professor of Logic in the University of Oxford until 1992. He was then without question the most distinguished and authoritative philosopher occupying an academic post in this country, and his productivity is undiminished since his retirement. His great series of books on Frege, his own independent contribution to the philosophy of language and to the philosophy of mathematics, his work in logic and in metaphysics and, finally, his observations on the nature of philosophy itself, constitute a body of theory unsurpassed in quality by the work of any of his contemporaries in the English-speaking world. The work is distinguished throughout by its originality, by its profundity and range, and by an unusually high level of intellectual unity. Its importance is universally acknowledged in the philosophical world, and its influence has been, and will be, great.

Dummett published his philosophical ideas in eight books and nearly a hundred articles, including one book and three articles in mathematical logic. His philosophical publications have given rise to a vast literature of responses, including the critical essays in that ultimate accolade—a volume in the *Library of Living Philosophers* devoted to *The Philosophy of Michael Dummett*.

Other interests

Anti-racism and support for immigrants

As has been described, Dummett, with his wife Ann, played an important role in combatting racism in Britain. As well as his work as an activist, he wrote or co-wrote five pamphlets and an article on this topic, and a book, *On Immigration and Refugees*.[106] Dummett's *Library of Living Philosophers* volume contains an article on 'Work against racism' by Ann Dummett, and on 'Immigrants and refugees: individualism and the moral status of strangers' by Kwame Anthony Appiah.[107]

[106] M. A. E. Dummett, *On Immigration and Refugees* (London, 2001).
[107] A. Dummett, 'Work against racism', in Auxier and Hahn, *The Philosophy of Michael Dummett*,

Voting systems

He had a strong interest in voting systems and published significant work on this topic, both theoretical and practical, in the form of two books and three articles, including an influential joint paper with Robin Farquharson, 'Stability in voting'.[108] He advocated the Borda count, and was able to put his views into practice when as Sub-Warden of All Souls (1974–76), he presided over the election of a new Warden (in which Patrick Neill was elected). After re-election to the British Academy, in 1995, he proposed a system of voting for the Philosophy Section to elect new members that allowed for weighting of negative as well as positive preferences, which was tried out in one election, alongside the existing system, but not adopted after objections that it was unrealistic to expect voters to have a preference, for each pair of candidates, between electing both and electing neither. Dummett's *Library of Living Philosophers* volume contains an article on 'Michael Dummett on social choice and voting' by Maurice Salles, who concludes, 'It is very difficult to convey the richness of the contribution of Michael Dummett to social choice and voting theory and to the practical voting procedures. [...] In French "hobby" is *violon d'Ingres*. I do not know whether Ingres, surely a great painter, was a good violinist. However, I am sure that Michael Dummett is a great social choice theorist.'[109]

Tarot cards and their uses

He pursued a passionate side interest in the games played with tarot cards, and in the cards themselves, about which he published six books and nearly forty articles. His researches are credited with establishing that the tarot cards are not relics of Ancient Egypt (as some like to believe), but originated in fourteenth-century Italy and fifteenth-century France. He also co-authored two books on the vogue for using Tarot cards for fortune-telling, *A Wicked Pack of Cards: Origins of the Occult Tarot* with Ronald Decker and Thierry Depaulis, and *A History of the Occult Tarot 1870-1970* with Ronald Decker.[110] Dummett's *Library of Living Philosophers* volume contains an article by Thierry Depaulis, 'The first golden age of the Tarot in France'.[111]

---

pp. 845–55; K. A. Appiah, 'Immigrants and refugees: individualism and the moral status of strangers', in Auxier and Hahn, *The Philosophy of Michael Dummett*, pp. 825–40.

[108] M. A. E. Dummett and R. Farquharson, 'Stability in voting', *Econometrica*, 29, 33–43.

[109] M. Salles, 'Michael Dummett on social choice and voting', in Auxier and Hahn, *The Philosophy of Michael Dummett*, p. 815.

[110] R. Decker, T. Depaulis and M. A. E. Dummett, *A Wicked Pack of Cards: the Origins of the Occult Tarot* (London, 1996); R. Decker and M. A. E. Dummett, *A History of the Occult Tarot* (London, 2013).

[111] T. Depaulis, 'The first golden age of the Tarot in France', in Auxier and Hahn, *The Philosophy of Michael Dummett*, pp. 901–12.

Yet further interests

Dummett was among the first to join the British Committee for the Reunification of the Parthenon Marbles, when it was established in 1983. He published two articles on the morality, or rather immorality, of nuclear deterrence, in 1984 and 1986.[112] In 1993 he published a practical book on English grammar and style, motivated by his experience of being a Finals examiner at Oxford shortly before his retirement.[113] He loved the Blues, a love ignited by hearing Bessie Smith's recording of 'Thinking Blues' in a record shop during his first trip to the United States, in 1955–6, and was proud to have heard Billie Holiday sing in a small bar on the South Side of Chicago in 1956.[114] (He listed 'listening to the blues', along with 'investigating the history of card games', in his *Who's Who* entry under Recreations.)

Appendix

As remarked above, Dummett (especially in his early papers) expected his readers to fill in the details of his deductions. We take this opportunity to put on record the intended completion of one well-known early argument, attested to in discussion with Ian Rumfitt around 1992.

Dummett's 1964 paper 'Bringing about the past' includes an extended comparison between two arguments.[115] The first is addressed to a parent who hears on the radio that a ship on which his son was sailing has sunk, and who prays that he was among the passengers rescued by another boat: 'Either your son has drowned or he has not. If he has drowned, then your prayer will not (cannot) be answered. If he has not drowned, your prayer is superfluous. So in either case your prayer is pointless.' The second is the apparently parallel argument for fatalism: 'Either you are going to be killed by a bomb or you are not going to be. If you are, then any precautions you take will be ineffective. If you are not, all precautions you take are superfluous. Therefore it is pointless to take precautions.' In the article, Dummett contends that both arguments are fallacious. In the case of the second argument, Dummett allows that the fatalist may infer from 'You will be killed' to '(Even) if you take precautions, you will be killed' and from 'You will not be killed' to '(Even) if you do not take precautions,

[112] M. A. E. Dummett, 'Nuclear warfare', in N. Blake and K. Pole (eds.), *Objections to Nuclear Defence* (London, 1984), pp. 28–40; M. A. E. Dummett, 'The morality of deterrence', *Canadian Journal of Philosophy* supplementary vol. 12 (1986), 111–27.

[113] M. A. E. Dummett, *Grammar & Style for Examination Candidates and Others* (London, 1993).

[114] Dummett, 'Intellectual autobiography', p. 13.

[115] Dummett, 'Bringing about the past'.

you will not be killed'. However, he deems it 'clear' that, on any use of '(even) if' on which this inference is valid, it is impermissible to pass from this last conditional to the fatalist's conclusion that 'Your taking precautions will not be effective in preventing your death.'[116]

Many readers have found this claim far from clear. Dummett understands 'Your precautions will be effective' as tantamount to the conjunction of two conditionals: 'If you take precautions, you will not be killed' and 'If you do not take precautions, you will be killed'. Using the letters 'P' and 'K' to symbolise 'You will take precautions' and 'You will be killed', his claim is that where '$\rightarrow$' is any conditional operator which validates the schema

$$(*) \qquad A \vdash B \rightarrow A,$$

the conditionals $\neg P \rightarrow \neg K$, $P \rightarrow \neg K$, and $\neg P \rightarrow K$ form a consistent triad.

There is, in fact, a strong argument for this claim. One instance of (*) is $\neg K \vdash P \rightarrow \neg K$; another is $\neg K \vdash \neg P \rightarrow \neg K$. Thus, if we had $\neg P \rightarrow \neg K$, $P \rightarrow \neg K$, $\neg P \rightarrow K \vdash$, we should also have $\neg K$, $\neg P \rightarrow K \vdash$, whence $\neg P \rightarrow K \vdash K$. Since (*) also yields $K \vdash \neg P \rightarrow K$, the hypothesis that $\neg P \rightarrow \neg K$, $P \rightarrow \neg K$, and $\neg P \rightarrow K$ are inconsistent generates the absurd result that $\neg P \rightarrow K$ is logically equivalent to its consequent. (As stated, the argument assumes that negation behaves classically. If it behaves intuitionistically, we reach the equally absurd conclusion that $\neg P \rightarrow K$ is equivalent to the double negation of its consequent.) Hence, on any use of '(even) if' which validates (*), the inference from '(Even) if you do not take precautions, you will not be killed' to 'Your precautions will not be effective' is fallacious, as indeed is the corresponding inference from '(Even) if you do not pray, your son will have been rescued' to 'Your prayer makes no difference'. Although he does not spell this out in the published article, Dummett confirmed that he had precisely this argument in mind.

Acknowledgements

Daniel Isaacson is grateful for help in writing about Michael Dummett from Michael's daughter Suzie Dummett, from Suzanne Foster, the Archivist at Winchester College, from Alex May, Research Editor at *The Oxford Dictionary of National Biography*, from Peter Brown and James Rivington at the British Academy, and from William Abraham, John Crossley, Helen Lauer, John Lucas, Paolo Mancosu, Brian McGuinness, Christopher Peacocke, Mark Rowe, Dana Scott, Michael Screech, Timothy Smiley, Göran Sundholm, William Tait, Kenneth Wachter and David Wiggins; he is grateful to Oxford University Press for permission to reuse some passages from his entry on Michael Dummett in the *Oxford Dictionary of National*

---

[116] Dummett, *Truth and Other Enigmas*, p. 341.

Biography,[117] and to Walter de Gruyter Publishers for similar permission with respect to his Biographical Sketch of Michael Dummett in, *Truth, Meaning, Justification, and Reality: Themes from Dummett*.[118] Ian Rumfitt thanks Christopher Peacocke for his comments on the assessment of Dummett's philosophical work. Both of us are grateful to Sir John Vickers, the Warden of All Souls, for giving us access to the college's file on Michael Dummett.

Note on the authors: Dr Daniel Isaacson is Emeritus University Lecturer in the Philosophy of Mathematics, and Emeritus Fellow of Wolfson College, Oxford. Professor Ian Rumfitt is Senior Research Fellow, All Souls College, Oxford; he was elected a Fellow of the British Academy in 2018.

---

[117] D. Isaacson, 'Dummett, Sir Michael Anthony Eardley (1925-2011)', *Oxford Dictionary of National Biography*, https://doi.org/10.1093/ref:odnb/104464 (accessed 7 September 2018).

[118] D. Isaacson, 'Michael Anthony Eardley Dummett: a biographical sketch', in M. Frauchiger (ed.), *Truth, Meaning, Justification and Reality: Themes from Dummett* (Berlin, 2017), pp. 1–12.

Alan Douglas Edward Cameron

13 March 1938 – 31 July 2017

elected Fellow of the British Academy 1975

by

ROGER S. BAGNALL

Fellow of the Academy

Biographical Memoirs of Fellows of the British Academy, XVII, 229–246
Posted 9 November 2018. © British Academy 2018.

ALAN CAMERON

Alan D. E. Cameron died on 31 July 2017. He was one of the leading scholars of the literature and history of the later Roman world and at the same time a wide-ranging classical philologist whose work encompassed above all the Greek and Latin poetic tradition from Hellenistic to Byzantine times but also aspects of late antique art.

He was born at Windsor on 13 March 1938, the son of Douglas and Bertha Cameron. His father's family originated in the Scottish Highlands village of Culbokie, north of Inverness. Douglas Cameron was in the insurance business and Bertha a housewife. Alan grew up in Egham, Surrey, and in 1946, along with his lifelong friend John North, entered Colet Court, the preparatory school for St Paul's School, where his father had been a pupil (and where his younger brother Geoffrey also studied, as would Alan's son Daniel in his turn). Although his parents did not have intellectual interests, they were supportive of their son's precocious academic gifts and let him live in a 'shed' in their back garden, where he had room for his books and peaceful conditions for study; Geoffrey describes it as a 'semi-permanent chalet construction'. He learned classical languages, begun at Colet Court, very quickly, and his academic bent was already visible in these preparatory years, as can be seen from an episode that I owe to Michael Yudkin, another classmate at Colet Court: 'The stocky and powerful Mathematics master, Mr Robinson, was in charge of games pitches. When the pitches were soggy, Mr Robinson used to bellow a series of set phrases to warn boys off using them: "You don't walk across that field"; "May I remind you that you don't walk across that field"; "You seem to have forgotten: you don't walk across that field".' Alan and Martin West turned these phrases into a Greek jingle, set to a tune, which went like this:

οὐ διαβαίνεις τὸν ἀγρόν.
οὐ διαβαίνεις τὸν ἀγρόν.
δύναμαι μιμνήσκειν σε, φαίνει γὰρ ἐπιλελησόμενός, σ' οὐ διαβαίνοντα τὸν ἀγρόν;'

But his more everyday concerns and engagement with the non-academic world are also already visible in a surviving diary he kept towards the end of his fifteenth year, from 1 January until 12 April 1953. In it, food, television, cinema, dental work, weather, shopping, card games, housework, rowing, trains and buses, astronomical observations (and even tea with Lady Herschel[1]), chapel, confirmation classes, and much else figure prominently alongside homework and classes, which he describes

---

[1] The widow (née Catherine Margaret Browell) of the third Baronet Herschel, who was the great-grandson of the famous astronomer Sir William Herschel, after whom the astronomy society of which Alan was a founder was named, see below.

without much interest or emotion.[2] A day ill at home was occupied by Agatha Christie and plays on the radio. In later life he disliked organised religion (he records his first communion on 20 March) but remained addicted to television; he was renowned for writing scholarly articles on a yellow legal pad while lying on the couch watching programmes such as *Star Trek* or *Perry Mason* on the screen.

Alan was a student at St Paul's from 1951 to 1956, commuting to Hammersmith from home. In his final year he won five prizes in Classics. These years, marked both by the teaching of W. W. Cruickshank for Latin and E. P. C. Cotter for Greek, and the friendship of Alan's classmates Martin West and John North, were decisive in his formation.[3] As he later remarked, 'Like all British classics students half a century ago,[4] we spent what now seems an extraordinary, not to say disproportionate amount of our time on verse composition, in both Latin and Greek.' Alan acknowledged Martin West's superior gifts in this domain, but he himself taught verse composition informally at Columbia University for decades. Despite the very narrow curriculum of St Paul's, however, it was also there that Alan was introduced, in 1951, by another of the masters, M. S. McIntosh, to the Greek epigrams of the Greek Anthology, which was to play a critical role throughout his scholarly life.[5]

From St Paul's, after a very brief stint in the army,[6] and the better part of a year enjoyably spent teaching Latin at Brunswick School near Brighton in Sussex, he gained a scholarship in 1957 at New College, Oxford, where his tutors were Eric Yorke and Geoffrey de Ste Croix. He obtained Firsts in Mods in 1959 and Greats in 1961; undergraduate prizes included the Craven Scholarship, the De Paravicini Scholarship and the Chancellor's Prize for Latin Prose. His contemporaries, several of whom were

---

[2] He did indeed later claim (to Charlotte Innes) that he was not much of a student at this age, although the academic record contradicts this recollection.

[3] See Cameron's remarks in 'Three tributes given by Jane Lightfoot, Alan Cameron and Robert Parker in memory of Martin Litchfield West', All Souls College, Oxford, 24 October 2015. The three classmates took part in producing the privately published *Apodosis: Essays Presented to Dr W. W. Cruickshank to Mark his Eightieth Birthday* (1992). Cruickshank was the more scholarly of the two Classics teachers, the sociable Cotter being best known for his books on bridge; see on him M. L. West in his Balzan Prize acceptance speech, printed in P. J. Finglass, C. Collard and N. J. Richardson (eds.), *Hesperos: Studies in Ancient Greek Poetry Presented to M. L. West on his Seventieth Birthday* (Oxford, 2007), p. xx. Although Cameron strongly emphasises Cruickshank's role in his references to St Paul's, John North has confirmed Cotter's importance as a teacher. It was Cruickshank, however, who formed a kind of external conscience for some of the students later on (a point I owe to Stephanie West).

[4] A statement true only of elite boys' schools, actually.

[5] See Cameron's *Wandering Poets* (Oxford, 2016), p. 10, on which a number of points in the next couple of paragraphs depend. McIntosh, according to John North, was the form master in their first year.

[6] Which he chose to do at this point, deferring university, and enjoyed, getting on well with the cross-section of men undergoing military training. But he was discharged after six weeks because of the hereditary knee ailment Osgood-Schlatter disease.

also members of Eduard Fraenkel's seminar,[7] included Martin West, Christopher Jones, Stephanie West and Averil Sutton, who became his first wife. Alan has written, perhaps with hindsight, about his chafing against the classical Oxford of the 1950s, with its curricular rigidities and separation of literature from history, although he acknowledges the important role that Fraenkel's seminar played in his scholarly development.[8] His relationship with Oxford, despite the easy brilliance of his undergraduate career, seems to have remained somewhat ambivalent. He neither pursued a further degree there, nor sought a fellowship there (there are various stories about that). As he later said, 'If I had done research in Oxford, I might never have turned to late antiquity.'[9]

But his turn towards the later empire was driven as well by more positive stimuli, most directly from reading (along with Averil) a copy of Gibbon he bought shortly after his Oxford finals and took on a Black Forest holiday; perhaps more indirectly from the burgeoning of interest in the period (and specifically in late paganism), with developments such as the appearance (in 1963) of *The Conflict between Paganism and Christianity in the Fourth Century*, edited by Arnaldo Momigliano, who became an important mentor.[10] The late Roman focus and work on Claudian, however, were clearly taking shape already before the publication of that book.

Immediately after Greats, he took up a post already in hand, on Yorke's recommendation, before he took his finals (Assistant Lecturer, then Lecturer, in Humanity, i.e. Latin) at Glasgow, where he spent three years (1961–1964). The teaching must have been somewhat strange for someone with Alan's background. The Professor of Humanity, the influential and eccentric C. J. Fordyce, expected the junior members of staff to do all the correcting of papers in Latin composition, feeding him lists of howlers he could use in class.

Glasgow had, however, an excellent copyright library, and already during those years, in 1963 and 1964, he began the writing of the torrent of publications that marked the next half-century, first with some short notes and then with longer articles, on (among other things) Ammianus, the *Historia Augusta*, Palladas, and 'Christianity and tradition in the historiography of the late Empire',[11] this last jointly with Averil Cameron, whom he had married after his first year at Glasgow, when she graduated

---

[7] See S. West, 'Eduard Fraenkel recalled', in C. Stray (ed.), *Oxford Classics: Teaching and Learning 1800-2000* (London, 2007), pp. 203–18. Fraenkel had retired from the Corpus Chair of Latin in 1953 but continued his seminars as well as his lectures.

[8] A. Cameron, *Callimachus and His Critics* (Princeton, NJ, 2007), p. x.

[9] Cameron, 'Claudian revisited', in his *Wandering Poets*, p. 134.

[10] This is a suggestion of C. P. Jones in his remarks at the memorial service at Columbia. Momigliano figures in the acknowledgments in *Claudian* and *The Greek Anthology*.

[11] Averil Cameron and Alan Cameron, 'Christianity and tradition in the historiography of the Late Empire', *Classical Quarterly*, n.s. 14 (1964), 316–28.

from Oxford. She was given a graduate scholarship from Glasgow and began work there on the dissertation that was to become her book on Agathias, a topic suggested by Robert Browning. One can see in Alan's early articles the outlines of preoccupations that would remain with him to the end of his life; his last article on Palladas dates to 2016. And the characteristics of his scholarly writing seem fully formed almost from the start: the vast command of ancient literature and modern scholarship, the philological precision in reading texts, the taste for polemic, the self-assuredness, the crisp and fluent prose, the wide scholarly network.

He was already, since leaving Oxford, engaged with Claudian, the Greek Alexandrian poet known for his Latin panegyrics, and he gave his first public lecture (to the Roman Society, in 1963) on the poet. As he says, 'I was looking for a topic that combined Latin and Greek and was both literary and historical. All my interests seemed to come together in Claudian: a brilliant Latin poet; a Greek by birth and also a Greek poet; and a major but unexploited source for an important but otherwise ill-documented period of history.' He took on the subject 'with the self-confidence of which only twenty-two-year-olds are capable' in the face of contrary advice from 'Tom Brown' Stevens.[12] His landmark article 'Wandering Poets: a literary movement in Byzantine Egypt',[13] which appeared in 1965, was part of this same move into the poetry of the late empire, both Greek and Latin.

In *Claudian: Poetry and Propaganda at the Court of Honorius* (Oxford, 1970), we find his distinctive qualities as a scholar and writer already on full display at length. Polemic, for one, in the preface: 'My acknowledgments to published works are fully recorded in the annotation—often, alas, by way of rebuttal. I could wish that this had not been necessary, but where my views differed widely from those generally held, it would have been misleading to state them without full justification.' The footnotes indeed cite the secondary literature widely, but not so as to overwhelm. Finding primary sources missed by previous scholars plays a larger role, as does sheer logical argument, albeit reinforced by forceful assertion. Taking a strongly chronological approach to the poems, he described Claudian as a propagandist for Stilicho. He also pointed to Claudian's ability to occupy an important role at a Christian court despite being a pagan. One sees here a foreshadowing of themes prominent in later work, particularly *The Last Pagans of Rome* (Oxford, 2011), rejecting views that saw Christians and pagans as separate, antagonistic groups.

In 1964 he was appointed as lecturer in Latin at Bedford College London. Like many (but far from all) of his contemporaries, he had never considered pursuing a

---

[12] Cameron, 'Claudian revisited', p. 134.
[13] A. Cameron, 'Wandering poets: a literary movement in Byzantine Egypt', *Historia: Zeitschrfit für Alte Geschichte*, 14 (1965), 470–509.

doctoral degree, and in later life he took pleasure in correcting anyone who addressed him as 'Doctor' Cameron.[14] But in 1971, the year after the appearance of the book on Claudian, he was promoted to Reader at Bedford, and in 1972 appointed to the chair in Latin at King's College London. It is impossible to imagine such a trajectory today. Already while at Bedford he and Averil spent a year visiting at Columbia University, both of them teaching in the graduate school, caring for their newborn son Daniel, despite Gilbert Highet's best attempts to dissuade them from coming with an infant, and chancing to be there in the most tumultuous year in that university's modern history, with the student uprising of spring 1968.[15] Somehow, that experience did not put him off New York, a city he loved; indeed, he found the events exciting and stimulating. He returned to New York in 1977, when he received and accepted the offer of a permanent appointment at Columbia as Anthon Professor of Latin Language and Literature, a post he held until his retirement in 2008. He became very much at home in American culture and (Democratic) politics.[16]

Along with all his other projects, the London years were marked by a deep involvement in the large British Academy-funded project the *Prosopography of the Later Roman Empire*, for the first volume of which (Oxford, 1971) he wrote most of the literary entries; he was part of the editorial team for the second volume (Oxford, 1980) and chaired it in his last year in London. This involvement certainly contributed to his historical side and his involvement with material culture.

The five years at King's College were anything but quiet; what was to become later *The Greek Anthology* (and give birth earlier to *Porphyrius* and *Circus Factions*) was under way, under 'the provisional title *Early Byzantine Epigrams in the Greek Anthology*'.[17] Here we see already the accidental (or opportunistic) character that Alan accurately ascribed to many of his books, for he goes on to say, 'While writing it [a chapter of his planned book] I learned that Louis Robert had identified some of the epigrams in question on a statue base recently excavated in Istanbul. To my surprise and delight he replied to my inquiries by inviting me to publish the monument in his stead.' Fearsome polemicist though Robert could be, he was also capable of great generosity and kindness towards young scholars, as I experienced myself.

This side road led to two books, one directly focused on the epigrams that were its genesis, *Porphyrius the Charioteer* (Oxford, 1973), and the second and more directly

---

[14] Averil Cameron, instead, moved her doctoral supervision to Arnaldo Momigliano in London, as Glasgow would not permit her to continue in absentia.

[15] See his 'Student rebellion at Columbia', *Oxford Magazine* (Trinity no. 8, 1968), 403–4.

[16] In hospital, coming out of anaesthesia after his last operation in summer 2017, on being asked the usual question (to test his mental condition) about who was president of the United States, he replied 'I prefer not to say.'

[17] A. Cameron, *Porphyrius, the Charioteer* (Oxford, 1973), p. v.

historical *Circus Factions: Blues and Greens at Rome and Byzantium* (Oxford, 1976). *Porphyrius* begins with a publication of a statue base bearing epigrams already known from the Greek Anthology honouring a famous charioteer of this name, before the book goes on to 'investigate a number of wider problems', which Alan admits 'may at first sight seem to have little to do with either Porphyrius or his monument'. Indeed, the book proceeds to try to reconstruct the entire set of monuments that would have borne the remainder of the many epigrams concerning Porphyrius and other chariot-eers, and to establish the critical text and chronology of all of these; the texts are carefully collated against the various manuscripts of Planudes' collection of epigrams. Concerns and methods of analysis found in the later book on the Anthology are visible here; but so too is an early engagement with the archaeological history of Constantinople and with the artistic side of the monument, as the reliefs are investigated in considerable detail and with the usual seemingly exhaustive command of the bibliography in all languages. The interest in Constantinople is also reflected in a seminar jointly led with Averil Cameron at King's College in 1974–76, which led eventually to a volume translating and commenting on an eighth-century text, the *Parastaseis Syntomoi Chronikai*, edited by Averil Cameron and Judith Herrin, with Alan listed as a contributor.[18]

For all this technical learning, however, the book is not limited to these detailed (and, as he described *The Greek Anthology*, 'austere') studies. It begins the discussion to be continued in *Circus Factions* of the world of the charioteers. He notes, 'The Byzantines had two heroes, Norman Baynes once remarked: "the winner in the chariot race and the ascetic saint". There is a whole literature on the ascetic saint, yet not so much as a single good article devoted to the fame of the charioteer.' It must be said that forty-five years later there is a bit more bibliography on the circus, but the ascetics' lead in scholarly literature has only widened to proportions unimaginable in 1973. Alan Cameron, a sports fan (baseball and tennis, to be sure, not chariot racing) and no friend to organised religion, did not contribute to that development.

The conclusion to *Porphyrius* begins the work of providing a larger context to the charioteer's monuments: at least a temporary discontinuance around the year 500 of wild-beast shows and pantomimes, and a rise in factional violence during the reign of Anastasius. The statues of Porphyrius, unprecedented as far as we know, are to be explained 'as part of Anastasius' wider policy towards the factions', of trying to keep the Blues and Greens fairly evenly balanced, with Porphyrius changing faction fairly frequently.

---

[18] Averil Cameron and J. Herrin (eds.), *Constantinople in the Early Eighth Century: the Parastaseis Syntomoi Chronikai*, in conjunction with Alan Cameron, Robin Cormack and Charlotte Roueché (Leiden, 1984).

Circus Factions, the 'companion' volume to *Porphyrius*, opens with a declaration of war on the orthodox interpretation of the factions in 'social, religious and political rather than sporting terms' and the consequent focus of scholarly investigation on questions related to these domains: 'The most obvious and important aspects of the subject have never been studied at all.' The most recent synthetic work, published only in 1968, J. Jarry's *Hérésies et factions dans l'Empire Byzantin du IVe au VIIe siècle*, is dismissed as a 'spectacular marriage of traditional falsehood with original fantasy' that has 'put it beyond the reach of ordinary criticism'. The first part of *Circus Factions*, we are told, is devoted to demolishing previous scholarship, despite which 'I have silently ignored most of the wilder flights of my predecessors'. The second part is then Alan's own construction of a 'realistic account' of the phenomenon. Along the way, he has found it necessary to investigate 'another underresearched topic, the history of popular entertainment in the Roman world'. The preface ends the acknowledgments with 'The argumentation and presentation of the whole book owes most of such lucidity as it possesses to the vigilant criticism and unfailing judgement of Averil Cameron, who also removed most of its adverbs.'

The challenges offered by the subject were in fact considerable, as Alan set out not merely to study the circus factions (an inaccurate description, he argues, but retained because of its wide usage) at their peak in the early Byzantine period, but over a period of about 1,200 years from the early Roman empire to the twelfth century; and, therefore, not merely at Byzantium but in the Roman world as a whole. This, he says, has never been attempted: classical scholars have stuck to the high empire, Byzantinists to later Constantinople, the two agreeing that there was a radical change between the earlier and later factions, a change representing more fundamental changes in the Empire, whereby the factions became in effect political parties representing the will of the people. Probably no one familiar with Alan's scholarly modus operandi even at this stage of his career will be surprised to learn that he discards all of this as so much rubbish. The supposed differences between early and late empire are distilled into six points, of which three are 'simply false', while the others reflect Byzantinists' ignorance of the early empire. Not that there was no change, but the real changes that occurred bear no resemblance to the traditional picture. 'The assumption which I wish particularly to combat is that these changes represent a growth of popular sovereignty.'

The argument is long and complex, and I shall not summarise it here, but it begins (characteristically) with an act of clarification. The *factio*, properly speaking, denoted the professional management and staff of the racing groups. It does not refer to the much larger group of partisans. And neither of these was responsible for financing the races. Constantinople had no 'demes' in the Athenian sense. Only with these (and other) distinctions clear can the rest of the discussion rest on a solid footing. Like

Porphyrius, *Circus Factions* displays Alan's usual vast command of both primary sources and secondary literature, in this case over a breathtaking span in time and space. Unlike many of his other books, it has hardly any literary side, that having been reserved for *Porphyrius*; the work is essentially historical and its method largely philological. Previous scholars have either not read the sources or have failed to read them accurately. No quarter is given.

Despite the philological detail, the book remains highly readable. Alan sent an inscribed copy to Charlton Heston, along with a printed lecture ('Bread and Circuses'), thinking the actor's experience in driving a chariot in *Ben-Hur* might make it of interest to him. Heston's charming thank-you letter (6 April 1977) says 'Both would have been useful to me when I was acquiring my limited competence as a charioteer.' He suggests having lunch with Alan on his next visit to London; Alan did eventually lunch with his agent but did not follow up beyond that.[19]

These first three books, along with his many articles, gave Alan an early reputation. There does not, indeed, seem to have been a time when his gifts went unrecognised; the invitation to Columbia for 1967/8 was an early sign of his international reputation. Although not quite as young at election as his (six months' senior) schoolmate Martin West (FBA 1973, at thirty-five years old, a near record), Alan was elected to the British Academy at thirty-seven (1975).

Given the rapid succession of books, it is perhaps not surprising that there were (at least by his standards) relatively few articles in the first half of the 1970s, and the early years after the move to Columbia were also not the most productive in that respect, as he adjusted to academic and domestic change and settled into a university still going through a profound financial crisis. He shortly found himself acting chairman of the Department of Greek and Latin, and then chairman. Administration was not his natural métier, to put it mildly, and Alan never changed in that respect any more than in others. But his scholarly stature gave him credibility with the administration, and he put his classically trained persuasive powers to good use.[20] He also put considerable effort into improving the quality of the department's faculty and graduate students, with considerable success. He did not, however, take much interest in his memberships in scholarly organisations, and his curriculum vitae is singularly bare of the kinds of professional service that most academics routinely undertake.

[19] I am indebted to Carla Asher for a copy of the letter, and to Charlotte Innes for the information on the limited follow-up.
[20] He was proud of (and I grateful for) what he regarded as his greatest success in persuasion, getting the dean to allow me to be brought up for tenure at a time when that was nearly impossible. But his gifts as a persuader go back to childhood; his brother records being conned into lending his cashbox for a penny a year to serve as the treasury of the astronomy society (see below).

The later 1970s were also in reality more productive on the scholarly front than would appear from a list of publications, as it was then that he wrote the core of his book *The Greek Anthology: from Meleager to Planudes* (Oxford, 1993), already under way in the early 1970s, as we have seen. This was indeed submitted to Oxford University Press in 1980, but 'languished in the limbo of copy-editing for a record decade, while I made fitful additions in the intervals of pursuing other projects'. The 'limbo' in question was actually Alan's desk, as his passive resistance to dealing with the copy-editor's queries made it easier for him to do almost anything else than come to terms with the minutiae of putting the book in final form. Only after prodding (from Debra Nails, he says in the preface) did he finally finish the job, dating the preface in April 1991. Whether eleven years in copy-editing is a record, I do not know, but it might be. Alan acknowledges McIntosh's impact on the thirteen-year-old Cameron, with Greek epigrams 'declaimed in a sonorous Irish brogue'. But he credits later work by Gow and Aubreton for leading him to try to 'penetrate the deeper mysteries of structure and sources', a good capsule description of Alan's interests in most of his books. With characteristic disconnection between printed polemic and personal relationships, he rejects 'Aubreton's methods and conclusions in their entirety' while thanking him for assistance.

The approach too is typical. After tracing the development of the epigram as a type of poetry, he looks at the collections of epigrams that formed the basis of the eventual (*c*. AD 900) work of Constantine Cephalas, and its later derivatives the *Palatine Anthology* (later in the tenth century) and the *Planudean Anthology* (1301), our two key manuscripts, along with some shorter extracts. The first of these earlier works, the *Garland* of Meleager (dated here *c*.100–90 BC), played a decisive role in creating (so Alan argues) the genre of anthology and its defining focus on short poems mainly in elegiac couplets. The second major anthology, that of Philip of Thessalonica, is argued to belong to the reign of Nero. In both cases, the book provides a detailed and incisive account of the poets and poems included, and how these are to be dated. The third major source in the eventual anthology was the *Cycle* of Agathias (*c*. 568). Cephalas' work, which does not survive, is tied to the great migration of classical texts from 'uncial' to minuscule script, and a detailed analysis teases out the condition of the copies that he had to work from and the way in which each anthology was organised, along with Cephalas' working methods and those of the creators of the surviving codex. Cephalas is the hero of the book, one might say, subsequently all but forgotten because of the fame of Planudes, whose anthology dominated until the Palatine Anthology was published in the nineteenth century.

We shared an interest in consuls, I from studying the chronological usages of the Egyptian papyrus documents and he from his knowledge of the social and political milieu of the late Roman elite. Consulates indeed already figure prominently in

Claudian. Out of a couple of small notes grew a joint project on the consuls, which eventually resulted in our *Consuls of the Later Roman Empire* (Atlanta, GA, 1987, with Klaas Worp and Seth Schwartz), to which Alan contributed much of the historical background on the consular elite and sections on the literary sources. But this work also led Alan in many other directions visible in his articles, among them late Roman aristocratic naming conventions and the consular diptychs. An interest in art as evidence for the aristocracy was not entirely new, and his detailed description of the reliefs on the Porphyrius monuments showed his engagement with the literature on late antique art, but it is really from 1981 on that he became deeply engaged with the ivories; at his death he left unfinished a planned work on the subject (jointly with Anthony Cutler). A collection of his articles on aspects of the art of Late Antiquity is currently in press.[21]

Another offspring from the 1980s, and a descendant again of *Claudian*, was the only other co-authored book in Alan's bibliography, *Barbarians and Politics at the Court of Arcadius* (Berekeley, CA, 1993). An interest in Synesius' *Egyptian Tale* announced in *Claudian*, but then laid aside, was brought back to life by reading a draft of an article by Tim Barnes in 1983: 'I was moved to strong disagreement and sent him a list of comments longer than his manuscript. But for that stimulus (and the lively exchange of views that followed, each of us convincing the other on some key points), this book would never have been written.' The joint authorship with Jacqueline Long and contributions from Lee Sherry were the result of a graduate seminar on Synesius that Alan conducted not long after.

The work of Synesius that had initially sparked Alan's interest, here referred to by its other title (*De providentia*), belongs to the limited body of evidence for the political situation in the court at Constantinople during the period after Theodosius' death (395) when Stilicho was regent in the West and Claudian writing his poetry. The book uses a detailed study of this work (of which a translation of the 'extraordinarily difficult Greek' is provided) and of Synesius' *De regno* to challenge most previous views of eastern politics in the period around 400, often seen as representing a contest between pro-barbarian and anti-barbarian parties, the barbarians in question being mainly the Goths. There have also been attempts to identify a pro-pagan 'party'. The set of views attacked here, however, 'rests entirely on a misinterpretation and misdating of Synesius's two works'.[22] Cameron and Long redate these works, and Synesius' ambassadorship from Cyrene to the court of Constantinople, two years earlier than they had usually been put; this may seem like a small matter, but it requires the veiled

---

[21] A. Cameron, *Historical Studies in Late Roman Art and Archaeology* (Leuven, in press), with an introduction by Jaś Elsner.

[22] A. Cameron and J. Long, *Barbarians and Politics at the Court of Arcadius* (Berkeley, CA, 1993), p. 9.

language of his works to be taken as referring to entirely different people and events and thus giving a completely divergent view of the period: 'since there was no *pro-barbarian* party, there was no antibarbarian *party*'. Synesius' writing is analysed in detail, 'revealing it to be a far more subtle, complex, and deceitful work than has been appreciated hitherto'. And Synesius himself 'was in fact an orthodox, if unconventional, Christian'. Thus, 'in short, there emerges an entirely new picture of the crisis of the year 400'.

Callimachus and his Critics (1995) is introduced with what by now is practically a topos, which is not to say it is untrue: 'Books (mine anyhow) have a way of growing in unexpected directions. This one started life as a reinterpretation of the *Aetia* prologue, its limited purpose to show that Callimachus's concern was elegy, not epic.' But it grew and grew: 'Much of the book is in fact more of a prolegomena to the study of Hellenistic (and so also Roman) poetry than a study of Callimachus alone. It is a social as much as a literary history of Greek poetry in the early third century.' That may leave some territory unclaimed, but not much! And, of course, he cannot pass up the occasion to point out that Callimachus' famous dictum ('A big book is a big evil' [or 'big bore' as Alan rendered it]) was tongue in cheek and can't be used to criticise the 524 pages devoted to him here.[23]

It will come as no surprise that the book opens with a bracing attack on much of what scholars have thought they could learn about the life, chronology, and character of the poet from his works and the scattered bits of ancient and Byzantine evidence. Callimachus emerges as a member of the Cyrenaean aristocracy, brought up at the Ptolemaic court, neither sycophant nor critic of Ptolemy Philadelphos. A highly realistic appreciation of the nature of early Hellenistic court life frames this discussion. The Cameron bulldozer proceeds through chapters' worth of clichés about Hellenistic poetry, leaving rubble in its wake. Cultural isolation, discontinuity, excessive learning, artificiality, marginality, remoteness from public life, and on and on, all are consigned to the trash heap. It is a 'widespread but unfounded modern notion that Hellenistic kings expected epics from their poets (Chs. X–XVI)', and Callimachus did not have to fight 'a life-long battle against epic poetry'.

As the remark quoted earlier suggests, the book is not only about Callimachus, although it does discuss all of his poetry in some detail. It is also about Theocritus, Posidippus, and so on. It is Callimachus in context. The book lacks a conclusion,

---

[23] Indeed, he argues that *biblion* cannot be taken to mean 'book' in our sense, referring as it does just to a papyrus roll: 'perhaps what Callimachus had in mind was the *sight* of a large roll in the hands of a poet about to recite—that sinking feeling we all know when we observe the thickness of the manuscript a lecturer takes out of his briefcase at the podium' (Cameron, *Callimachus*, p. 52). Alan was himself a superb lecturer, with a fine sense of audience and what it would listen to.

although one might quote a sentence from the last paragraph of Chapter 17: 'In effect, his polemic was a plea for originality and quality.'

'Like most of my books, this is not one I had planned to write.' Thus Alan found a slightly new phrasing of the usual disclaimer to describe the origins of his 2004 book on *Greek Mythography in the Roman World* (New York). It was in fact, he admits, a distraction from the long labour of writing *Last Pagans*, provoked by a paper sent him by Richard Tarrant on a mythographic work of the Roman period referred to as the *Narrationes*. Reading this, he noticed a similarity to passages in Callimachus' *Diegeseis*, which he had treated in his book on that poet, and he concluded that the *Narrationes* were 'a typical mythographic work of the early empire', designed to help make sure that the propertied classes were familiar with the 'stories every educated person was expected to know'. The emperor Tiberius was a devotee of mythographic trivia. As so often in Alan's work, it is the details of the process of transmission of information that are at the centre of the enquiry, certainly not mythology itself.

The retirement years, lived with characteristic vigour (he kept fit cycling and swimming), allowed Alan finally to finish *The Last Pagans of Rome* (New York, 2011), a massive volume of which the roots can be traced right to the beginning of his scholarly work and at which he worked for many years. At the same time, he produced a series of substantial and important articles sufficient to have earned anyone tenure, some of them side products of the great book but some in other familiar fields such as Palladas, consular diptychs, Ammianus, and the *Historia Augusta*: none of it irrelevant to *The Last Pagans*, of course, as indeed hardly anything he did truly was.

It would be a daunting task to summarise in any detail this capstone to a scholarly life. As Alan describes it, the project began to take shape three decades earlier but kept changing form and substance as it evolved. And yet its deep consistency with an entire career's work is obvious at every step. He sets out to dismantle the 'romantic myth' that the nobles of Rome were 'fearless champions of senatorial privilege, literature lovers, and aficionados of classical (especially Greek) culture as well as the traditional cults', when in fact they were 'arrogant philistine land-grabbers'. There was no pagan revival, there was no last stand of a pagan circle, the revolt of Eugenius was not about religion. The book sets out to demolish comprehensively almost everything usually claimed about the supposed conflict of Christianity and paganism, about the strength of paganism, about priesthoods. Christian rhetoric is (properly) treated as propaganda rather than fact. The list of supposed pagan authors is ruthlessly pruned. 'Many (too many) studies have been devoted to the religious beliefs of Rutilius' runs a characteristic sentence.[24]

---

[24] Cameron, *Last Pagans*, p. 207.

The path pursued to these conclusions is leisurely. An extraordinary and utterly original chapter is devoted to the origins of the term 'pagan' meaning 'non-Christian', which Alan sees as having been from the outset a neutral rather than hostile descriptor. He argues that it remains a useful term, dismissing the rival claims of 'polytheistic' and other terms as neither more accurate nor more neutral. The name, chronology, and works of Macrobius are treated at length. The claims for pagan aristocrats as editors of classical texts are systematically dismantled. The love of classical culture was shared by pagans and Christians, and not even the revival of interest in Silver Latin was specific to one religion. Chapters follow on correctors and critics, the revival of interest in Livy. Subscriptions in manuscripts are collected and studied, to refute five common assumptions (p. 422): 'that most of the subscribers were (1) pagans and (2) Roman aristocrats; (3) that the subscribers chose texts that both reflected and were intended to spread their pagan sympathies; (4) that they were consciously preserving precious pagan texts in danger of being lost; and finally (5) that they were performing some sort of serious editorial activity.' The books actually produced were not scholarly editions but luxury copies for the rich.

The claims made for the importance of Nicomachus Flavianus' *Annales* as a source for Zonaras, the *Historia Augusta*, Ammianus Marcellinus, and the *epitome de Caesaribus* is given a detailed refutation (64 pages). 'For the method used to "recover" Flavian's *Annales*, astrology would be a more appropriate analogy [than astronomy],' he concludes after a particularly devastating polemic (p. 628). The view of the *Historia Augusta* as part of a 'pagan reaction' is dismissed, and a date to 375–380 rather than the usual (since Dessau) *c*.395 is proposed. As for the work as pagan propaganda, 'The author of the *HA* was a frivolous, ignorant person with no agenda worthy of the name at all.'[25]

None of this came as a surprise to those who had followed the long arc of Alan's scholarly production. It is, in fact, striking how consistent was his set of interests and approaches from the start of his scholarly career to its end. He was always focused on solving problems, many of them straightforward matters like identification, dating, the sequence of events, the meaning of terms, and relationships between individuals and works. It is hard to see any development in his methods or style of argumentation, even though his knowledge of the sources and scholarly literature continued to deepen over the decades, hard though it might be for the reader of his early books to imagine any scope for such maturation. One book led to another, growing at an oblique angle, without formalised projects, and almost everything he wrote can to some degree be seen in embryo in *Claudian*. Even what for someone else would be a springboard to

---

[25] Ibid., p. 781.

broad synthesis always for Alan resolved itself into a set of problems to be solved and arguments to be won.

It is not as if a synthetic view of his subjects was missing from his thought. And he had a keen sense of realities and actual people of antiquity; they were not just objects of philological enquiry. Many of his arguments, throughout his writing, arise from a sense that some view must be wrong because it is incompatible with a broader understanding of the political, literary, institutional, linguistic, social, or religious context.[26] His more encompassing thoughts about these subjects can be found sprinkled throughout all of his books and articles. But he never sought to produce a synthesis in any of the subjects in which he was expert that would be non-argumentative in style and readily accessible for an audience that did not know six or more languages and was not prepared to follow him into every detail of a topic. His lecturing and teaching show that it was not inability to express himself in a less argumentative and technical way that led to this gap (as many of us would see it) in his work. Such writing was simply not what he enjoyed doing as a scholar, and his devotion to unenjoyed service work was not great. He also saw little value in theory (or Theory) and engaged with it only rarely and grudgingly.

Alan Cameron's American career was marked by the honours that one might anticipate, given his scholarly distinction: fellowships from the National Endowment for the Humanities, the Guggenheim Foundation, and the Institute for Advanced Study; election to the American Academy of Arts and Sciences (1978) shortly after his arrival in the United States, and later the American Philosophical Society (1992). *The Greek Anthology* received the Charles J. Goodwin Award from the American Philological Association, and *Greek Mythography in the Roman World* was awarded Columbia College's Lionel Trilling Award. The British Academy's own Kenyon Medal in 2013 capped his list of honours.

Alan was, as could readily be seen by all, enormously and justifiably confident of himself and his abilities. But this self-confidence, so visible in scholarly polemic and (not surprisingly) sometimes resented, generally translated in personal life not into arrogance but into complete comfort with others, whether in a classroom, lecturing to alumni, in social situations or conversing with staff in his apartment building. His large apartment on Riverside Drive near campus was the Classics Department's main space for social functions for decades, and his hospitality to guests—professional or personal, previously known to him or not, even unanticipated—is legendary, as is his generosity to, and enjoyment of the company of, graduate students and younger

---

[26] To pick an instance at random: *Circus Factions* p. 19 n. 1: 'Dvornik mistranslates the Balsamon passage "drew revenues *from* the entertainments for their upkeep". There was (of course) no revenue from ancient spectacles of this nature.'

scholars. His dislike for hierarchy and pomposity was noteworthy, perhaps a factor in his decision to make his career and life in the more fluid American environment. He rather cultivated the classic image of the absent-minded professor, constantly losing (but often later finding or having returned to him) all manner of things, a characteristic visible already at eleven years old, and failing to deliver grades and recommendations, not to speak of proofs, on schedule. (As department chair, I once fined him for failing to turn in grades.) On the other hand, despite his disengagement with all things administrative he had a good repertory of household skills and became expert in the use of the word-processing program Nota Bene for producing his books; he proudly reports in the preface to *Callimachus* that he submitted it in camera-ready copy. He was an intrepid user of the Thesaurus Linguae Graecae.

His scholarly work was unmistakably the centre of his life, and at times one suspected that he did little preparation for class, at least for undergraduate language classes; he liked lecturing better and put more effort into it, even if sometimes at the last moment. But his erudition, memory, clarity, wit and charm of exposition were such that he could teach almost any class on the spur of the moment and leave the students with a sense that they had learned something from a great intellect and had fun doing it. These qualities, coupled with a certain irreverence, made him an outstandingly successful lecturer on alumni tours and cruises, an activity in which he engaged often and took much pleasure.

It should also be remarked that Alan's self-confidence (in scholarly matters, at any rate), self-direction and unwillingness to do anything but what he wanted to do coexisted with two other characteristics central to his work. The first is a strong sense of what he owed to his teachers and informal mentors, difficult though it was at times not to think of him as the product of a scholarly virgin birth, given the lack of any formal research supervision at any point. In *Callimachus* he singles out Cruickshank, 'who at St Paul's School first introduced me to the meaning of scholarship'; Eduard Fraenkel; and Arnaldo Momigliano.[27] One might add the influence of Louis Robert to that list. The second was a deep collegial connection to both his contemporaries and younger scholars who influenced his work, and great scrupulousness in acknowledging these debts. Those mentioned in the dedications of *Claudian* and *Last Pagans*, thus spanning his career, included Tim Barnes, Glen Bowersock, Peter Brown, William Harris, Peter Knox, John Matthews, Momigliano, John North, and Martin West, but there were many others thanked in other books and articles. This rich network, including some of those he disagreed with in print, is reflected in his extensive archive of scholarly correspondence.

[27] Of these, only Cruickshank was still alive at the time *Callimachus* was published.

His ability to connect with all types of people owed much to the wide range of his interests, which can be traced from the Herschel Society, the astronomy club he, Martin West and Michael Yudkin founded at Colet Court,[28] with his brother Geoffrey joining at St Paul's (after Yudkin continued at another school), right down to his interests in rock and roll, film, opera, theatre, television, wrestling, baseball, and other aspects of popular culture; he loved to dance. His curiosity was vast, always tinged with a boyish enthusiasm, and he seemed unable to avoid becoming deeply knowledgeable and passionate about any subject that he went into at all. But his unpretentious and democratic manner kept all this from becoming as intimidating as it might have been.

Alan Cameron was married three times. Charlotte Innes, a writer, accompanied him on his move to New York in 1977 after she was accepted into the master's program in the Columbia School of Journalism, and became his second wife; that marriage ended in divorce. His third marriage, to a native New Yorker, the educator and university administrator Carla Asher, was happy and lasted nearly two decades until his death. He is survived by her and by his brother Geoffrey Cameron; his sister Sheila Hodge; his children Daniel and Sophie, from his marriage to Averil Cameron; and his grandson Silas, whom Alan was able to meet and enjoy towards the end of his life.

Acknowledgements
I first met Alan Cameron for lunch in London on 10 May 1977, when he had accepted the offer to move to Columbia University, and to a large degree can rely on my own observations for his forty years in New York. I have also drawn on reminiscences by the various speakers at the memorial services in New York and London. For supplementary information and insights, and above all guidance on his earlier years, I am indebted to Carla Asher, Averil Cameron, Geoffrey Cameron, Eleanor Dickey, Jaś Elsner, Alexander Garvie, Charlotte Innes, John North and Stephanie West.

Note on the author: Roger Bagnall is Professor of Ancient History and Leon Levy Director, Emeritus, Institute for the Study of the Ancient World, New York University, and Jay Professor of Greek and Latin and Professor of History, Emeritus, at Columbia University. He is a Corresponding Fellow of the British Academy.

---

[28] Cameron in West memorial (above, n. 3); memoir by Geoffrey Cameron.

Stewart Ross Sutherland

25 February 1941 – 29 January 2018

elected Fellow of the British Academy 1992

by

KEITH WARD

Fellow of the Academy

Biographical Memoirs of Fellows of the British Academy, XVII, 247–262
Posted 23 November 2018. © British Academy 2018.

STEWART SUTHERLAND

Stewart Ross Sutherland, Baron Sutherland of Houndwood, was a distinguished philosopher, an outstanding College Principal and University Vice-Chancellor and a notable public servant who sat on the cross-benches of the House of Lords from 2001, and was made a Knight of the Most Ancient and the Most Noble Order of the Thistle in 2002.

Stewart was born in Aberdeen on 25 February 1941, and was educated at Woodside School and Robert Gordon's College, Aberdeen. He took an MA in philosophy at Aberdeen University, gaining a First, and then gained another First Class degree, this time in Theology, at Corpus Christi College, Cambridge. The philosopher of religion Donald McKinnon was to be an important influence on his philosophical journey. While in Cambridge he married Sheena Robertson, whom he had met at the University of Aberdeen, where she studied medicine, then went on to have a career as a clinical virologist. They had three children, Fiona, who runs a gallery in Kent, Kirsten, who is a structural engineer, and Duncan, who is professor of nanoscience at the University of Aarhus in Denmark.

Stewart's first academic appointment was as a lecturer in philosophy at Bangor. At that time Welsh philosophy was much influenced by Wittgenstein (both for and against), and there was a strong interest in spelling out the distinctive and irreducible nature of religious language, as a set of language-uses which could only be coherently understood as part of a general form of life. This too was to be an important influence on Stewart's philosophical work. In 1968 he returned to Scotland as a philosophy lecturer at the newly established University of Stirling, later becoming a reader in the department. His main interests were in philosophy and literature, and he typically approached the traditional problems of philosophy by referring to important literary works. Dostoyevsky was a special interest, though Stewart did not hesitate to use detective stories too where it seemed appropriate. While in Stirling, he took an interest in Religious Studies, which was at first part of the philosophy department, and that interest was to influence his subsequent philosophical publications.

He was a Fellow in the Humanities Research Centre at the Australian National University in Canberra in 1974, and in 1977 he was appointed to the Chair of the History and Philosophy of Religion at King's College London, overseeing a Department of Religious Studies as well as contributing to the teaching of philosophy in the University and especially for the BD degree at King's College. His administrative skills quickly became apparent. He had an uncanny ability to chair meetings in a way that often defused arguments and led to new constructive proposals, most of which he had already carefully prepared before the meetings. He was, nevertheless, always ready to consider suggestions from others and incorporate them into the final decisions that were made.

Recognition of this skill led to his appointment in 1985 as Principal of King's College. There he initiated and oversaw major organisational changes, including mergers with Queen Elizabeth College and with Chelsea College, and with the School of Medicine and Dentistry. The latter merger helped to prepare the ground, some years later, for the further union with Guys and St Thomas's. The first Institute of Gerontology was founded at King's in 1986, marking what was to be a life-long interest for Stewart. With remarkable political skill, he oversaw the development of the Thameside Campus, taking over the lease on Cornwall House, and allowing the life sciences to be expanded and placed on a single site. He was thus a prime mover in enabling King's College to cement its place as a College with a truly international reputation. Before the mergers, King's College had between three and four thousand students. After them there were six thousand students, and now there are over thirty thousand. That is a remarkable development.

His administrative skills were noticed, and in 1990 he was appointed by the Crown to the Vice-Chancellorship of the University of London. This unique federation of nineteen very diverse institutions was going through a fissile and schismatic period, in which various colleges were asserting their independence and a new understanding of the federal nature of the University needed to be achieved. Stewart's voice of reasoned and persuasive influence was important in guiding this process, and he was able to pursue a number of projects of his own. He was a Governor of Birkbeck College from 1988-1991. He was particularly keen to stress the importance of including vocational training for teachers in the University—something that he was also to stress later in Edinburgh. And, while strongly supporting scientific research and innovation, he also insisted on the importance of the Humanities as essential parts of an education which could make provision for the ethical and political values which form the basis for a good personal life and a humane society.

The move to onerous and challenging administrative positions naturally affected his ability to publish, but the three volumes he did publish were important and original contributions to the field of philosophy, and to the philosophy of religion and the philosophy of literature in particular. He was the editor of the journal *Religious Studies* from 1986 to 1991, and he established it as the premier journal in the philosophy of religion. After his move to a primarily administrative and public career, he continued to publish papers and articles which, though they were short, were incisive and well argued, and constituted important defences of the value of a broad education which would include a stress on the humanities and on human values as well as on scientific research. He also managed to give no fewer than eleven series of named lectures at British Universities including the Wilde Lectures at Oxford in 1981–4, and was a Gifford Lecturer at Edinburgh in 2011.

He became Her Majesty's Chief Inspector of Schools in 1992, and in that capacity he was responsible for overseeing the formation of OFSTED (the Office for Standards in Education), which was a great improvement on the rather haphazard system of schools inspections that had then existed. He continued to argue publicly and forcefully for forms of education that would contribute to a personally fulfilling and socially cohesive life. In this, he continued the tradition of philosophy that was concerned with seeking what it is to be good as an individual, and what it is for a society to be ethically as well as materially rich. He showed that philosophy was not merely an abstract academic discipline, but an important resource for contributing reasonably, reflectively and humanely to major human institutions and to the public good.

In 1994 he moved to Edinburgh as Principal and Vice-Chancellor of the University of Edinburgh. There he instituted many significant developments, establishing a more secure financial structure, formulating a new staffing strategy and a restructuring of the University management, as well as a restructuring of the various curricula of University departments. These were thoroughgoing changes, and they were effected largely through a policy of choosing good people and letting them do their jobs without undue interference, while always making clear what he wanted done, and always being accessible and supportive.

He was successful in seeing that the University achieved excellent results in research excellence exercises, and also in enhancing attention to the quality of University teaching, as evidenced in a very positive institutional review by the Quality Assurance Agency. He convinced the University that there were strong educational reasons for including teacher education in a university context. To that end, he oversaw the merger of the Moray House College of Education with the University. This expressed his strong conviction that teaching practice should be based on a solid research foundation, and also that research-active institutions of higher education should be concerned to share their findings with schools to a greater extent. The title of one of his papers, 'The price of ignorance', given as the Hume Lecture in 1995 and published by the David Hume Institute, showed his belief that a strong foundation of both knowledge and values, laid at an early age, was a condition of a morally healthy and flourishing society.[1]

Stewart was also keen to expand the University's participation in relationships with other universities around the globe—for instance, with Stanford and with members of Universitas 21. He strengthened the already high reputation of Edinburgh for research in the sciences, overseeing the construction of new Medical School facilities, strengthening ties with Research Councils, and with such bodies as the Wellcome Millennial Clinical Research Facility and the National e-Science Centre, established jointly with the University of Glasgow. Stewart was the first Principal of the University for many years who was neither a scientist nor a medic, but he justified his appointment in full in that while he argued for the essential place of the Humanities in

[1] S. Sutherland, *The Price of Ignorance* (Edinburgh, 1996).

University education, he was wholly committed to promoting excellence in scientific and medical research.

In 2002–8 he was Provost of Gresham College, London, and in that capacity he introduced the practice of making videos of the lectures available on the world wide web. They have given these lectures and the College an international reputation for excellence. He also expanded the work of the College to make it a lively forum for debate in the heart of the City of London.

His combination of philosophical acumen and personal leadership skills led to his involvement with many fields of activity outside the University system. He served on the Board of the Higher Education Funding Council for England, on the Hong Kong University Grants Committee—for whom he conducted a major review of higher education—as Vice-Chair of the Committee of Vice-Chancellors and Principals (now Universities UK) and as Convener of the Committee of Scottish Higher Education Principals/Universities Scotland. The Secretary of State for Scotland appointed him as Chairman of the Committee on Criminal Appeals and Miscarriages of Justice Procedures. The work of this committee formed the basis of the 1995 legislation which introduced the Scottish Criminal Cases Review Commission.

In 1997 he was appointed as chair of the Royal Commission on Long Term Care of the Elderly. This recommended that all nursing and personal care should be provided by the government, either in care homes or if possible in their own homes. It also recommended that health and social care should be considered together and their budgets merged. The recommendations proved to be too radical at the time for the UK government. Yet the problem is increasingly obvious. The population of over-70s is predicted to reach 7–9 million by 2020, and Age Concern has characterised the situation of care for the elderly in England as unacceptable in a civilised society. The report of the Commission was examined in an independent review of free personal and nursing care in Scotland, in 2008, and the core proposals were implemented there.

These involvements in crucial issues in care of the elderly and in criminal justice were complemented by his membership of the Council for Science and Technology, and by his Chairmanship of YTL Education (UK), and of Frog Trade from 2013—both important enterprises concerned with educational standards and opportunities in Malaysia as well as throughout the world. He chaired the Associated Board of the Royal Schools of Music 2006–17, and was on the editorial board of the *Encyclopedia Britannica* from 2005, both appointments demonstrating his impressively wide range of interests. He was President of the Royal Institute of Philosophy 1989–92, and of the Society for the Study of Theology in 1985. He was for many years a member of the Goldsmith's Company, chairing their education committee, and becoming Prime Warden in 2012–13. The Company has founded ten Goldsmith's Sutherland

Scholarships for students from socially disadvantaged backgrounds to study philosophy of religion under the supervision of Pembroke College, Oxford.

Stewart was President of Alzheimer Scotland/Action on Dementia, President of the Saltire Society, President of the David Hume Institute, and of Scottish Care. His interests ranged over concern with the care of the elderly, with issues of criminal justice, with defending the importance and excellence of education in both the humanities and the sciences, with philosophy and theology, and with new problems and opportunities raised by advances in science and technology. To all of them he brought innovative ideas and firm and positive leadership.

In recognition of his work in so many diverse fields, he was knighted in 1995 and became one of the first fifteen new independent 'people's peers' appointed to the House of Lords in 2001, taking the title of Baron Sutherland of Houndwood after the Berwickshire home in which he took such delight. He has received honorary degrees from universities around the world, and holds honorary fellowships at King's College London, Corpus Christi College, Cambridge, and the University College of North Wales, Bangor. He was elected a Fellow of the British Academy in 1992 and a Fellow of the Royal Society of Edinburgh in 1995, being President 2002–5.

His intention was to return to academic research, and in particular to complete long-awaited books on world religions and on further thinking about the nature of religion in the modern world. It is no surprise that he had been unable to do this in the midst of his remarkably busy and successful administrative career. After his retirement, he continued to work hard on issues of special concern to him in the House of Lords, and unfortunately his illness and early death left his academic work uncompleted. Nevertheless, his contributions continued in a great number of occasional papers of substance and originality, and his published work remains important and relevant for younger scholars as they seek to address the problems and perspectives with which he dealt.

He was a philosopher with a main but not exclusive interest in religion, particularly in Christianity as a 'form of life' which offers a distinctive practical understanding of what it is to live well as a human being. He explicitly set out to provide a revisionary development of Christian tradition, believing that the legacy of Christian theism offered something of great value to society and to individuals.

He said himself that his revisionary view would probably not score very highly on the scale of orthodoxy. But he thought that the language and practices of Christian faith made possible a view of life and a way of living in the world that was distinctive and difficult, if not impossible, to express in any other way. Thus, his view seeks both to preserve a specific religious outlook and yet to revise that outlook in radically new ways. It neither defends a form of religious orthodoxy nor dismisses religion as false or irrelevant. What he writes is a significant contribution to thinking about the place of religion in the modern world, and an important contribution to rethinking the nature of religious faith.

His view is outlined in many papers and articles, but mainly in three books, *Atheism and the Rejection of God* (Oxford, 1977), *Faith and Ambiguity* (London, 1984), and *God, Jesus, and Belief* (Oxford, 1984). The first of these is an engagement with Dostoyevsky, especially with *The Brothers Karamazov*, and with the central conversation in that book of the brothers Ivan and Alyosha Karamazov about the reality of suffering and evil in the world.

This interest is central for Sutherland, for it illustrates the importance for him of literature as a vehicle of philosophical reflection, and also his conviction that any thinking about God must begin from a full acceptance that much suffering is real, morally unjustifiable and destructive of many traditional ideas of God.

Both of these convictions are controversial. Since the time of Plato, many have suspected that there is a war or at least a tension between poetry and literature on the one hand and philosophical speculation on the other. The conversation of the Karamazovs illustrates this well. It states the opposing views of both brothers, giving perhaps the most powerful argument against a good God in world literature, and leaving Alyosha without any obvious reply. Yet there is a sort of reply, in Alyosha's rejection of bitterness and rebellion, and insistence upon love of the beauty of the world and the cultivation of compassionate love. Can Alyosha's life be a reply to Ivan's arguments? Not, Sutherland suggests, in a purely intellectual or rational way. But perhaps the lesson here is that 'there is no single metaphysical picture' that can give a complete understanding of the world. That is the strength of great literature, that it provides no such coherent picture. It usually presents an ambiguous reality, a picture on the borderlands between belief and unbelief, where emotions can lie deeper than reasons, and where forms of life and ways of seeing the world are not based simply on the provision of publicly available and agreed reasons or evidence.

The grand metaphysical systems of the past, from neo-Platonism to the revised Aristotelianism of Aquinas and Hegelian Idealism, seem to many to have been dissolved by the sheer range and variety of specific forms of modern knowledge. Human minds have enough difficulty in mastering the small areas of research in which they are most interested. The task of forming a vast over-arching picture of reality into which all areas of research could be coherently fitted seems out of reach. Perhaps Hegel was right in suggesting that knowledge progresses in a dialectical fashion, by thesis countered by antithesis, as different aspects of human experience and knowledge continually interact with one another. But perhaps Hegel was wrong (as was Marx, in his version of Hegel-standing-on-his-head) in thinking that there was some super-rational synthesis into which this dialectic could be fitted. What we have are fragments of knowledge and belief, interacting indeed but never achieving 'the system', the absolute metaphysical truth.

This is very apparent in the work of Kierkegaard, and in works of literature, such as those of Shakespeare, Dostoyevsky and Goethe, which stress the plurality and diversity of human forms of life and ways of seeing the world, and the lack of any unifying synthesis that could resolve all these perspectives under one rational and intelligible system. Where there is scepticism about the ability of any one world-picture to be the obviously rational and coherent one, it will be impossible either to accept a world-view based on the natural sciences alone or one based on religious or philosophical considerations alone. This sort of deep metaphysical scepticism, which is yet allied to a serious search for truth in its various particular guises, lies at the heart of Sutherland's approach to philosophy.

The acceptance of ambiguity and lack of finality in our ultimate judgements about the world has an effect on what we take to be philosophical truth. In particular, theodicy becomes impossible, if that is thought to be the provision of good reasons that a personal God might have for creating or permitting horrendous suffering. Sutherland is clear that the idea of God as a wholly good person or even as an individual object, supernatural or natural, with whom one might have conversations or personal relationships, is untenable. It is totally incompatible, as Ivan claims, with the suffering of innocent children.

That is indeed going to require a revision to most traditional religious views, though it is probably nearer to sophisticated expositions of a traditional Thomist position than most people suspect. But the word 'God' still, he argues, has a distinctive use. That use is to make possible a view of the world *sub specie aeternitatis*. The phrase is perhaps best known because of its use by Spinoza, who wrote, 'Those things which are conceived as true or real we conceive under the form of eternity and their ideas involve the eternal and infinite essence of God' (*Ethics*, Part 5, prop.29, note). Again, 'eternity is the very essence of God, insofar as that essence involves necessary existence' (Prop. 30). Sutherland does not usually explicitly refer to Spinoza, and the idea that all things devolve by necessity from God, who contains the essential natures of all things, is not one with which Sutherland might be comfortable. Yet the ideas of non-temporality and of necessity seem to be implied, so that what is being said is that there is something non-temporal and non-finite (therefore not 'a thing') beyond contingent and transient existence. That is not another separate and distinct reality, but an aspect of this reality in which we exist, and an aspect which cannot be denied without the loss of sensitivity to important features of human existence.

However, for Sutherland this should not be thought of as an object or set of objects which we could contemplate or intuit. We do not, as Schleiermacher said, 'intuit the eternal'. Rather, we intuit this world in the light of eternity. Like Kant's regulative ideas, the idea of eternity does not correspond to a knowable reality. It enables us to see the world from a viewpoint which is not that of any human being or group of human beings. As Kant put it, it is like a 'focus imaginarius', an imagined 'absolute view of things as they are', beyond all partial and finite perspectives. It is ideal rather than real. As Sutherland puts it, it is a possibility rather than an actuality. Yet to appeal to it even as a humanly unrealisable ideal enables us to see the world in a distinctive way.

Nor is this just one possibility amongst others. It is (though he does not, I think, use this word) an authentic possibility. It enables us, and he does say this, to 'see our world as it is'. Thus, it is not just one option among others. It reveals an important truth about the world, which is that the world is more than just a collection of finite ultimately material entities. There is no appeal here to God as an extra or supernatural entity. It is admittedly difficult to see precisely what is meant by seeing the world from the perspective of eternity. One clue lies in a statement by Kierkegaard, to which Sutherland refers in *Faith and Ambiguity*, that eternal significance cannot be found in world-historical terms. It is 'inward' and involves a certain sort of passionate commitment. The real subject is the ethically existing subject, Kierkegaard writes. Such ethical existence requires self-knowledge and a purity of heart which provides the perspective of eternity on [one's] doing and deciding.

Like 'eternity', purity of heart is left undefined and perhaps indefinable, but it points to a state in which one's attention is not fragmented and scattered among a number of no doubt pressing temporal projects, but is concentrated on a 'transcendent' order beyond finite concerns, 'a transcendent order of eternal values', which are real but never completely grasped or embodied in temporally identifiable forms.

Sutherland's approach has been and is still of importance in philosophical enquiry. Philosophers such as John McDowell and David Wiggins have explored the possibility that values are objective features of reality, and have proposed a form of 'enriched naturalism' which does not invoke any sort of supernatural reality, but sees values as part of the natural world. They are reluctant to speak of God in this context, though the philosopher Fiona Ellis has argued that some ideas of God can be included as pointing to objective features of reality without deploying the idea of a 'supernatural' order of beings. This is very similar to Sutherland's position, which is sceptical of full-on metaphysics, yet does wish to see values, and the associated purposeful activity of pursuing values, as part of the furniture of the natural world, and thus as part of our ordinary and natural—not 'religious' in positing special forms of unusual experiences or unknowable types of supernatural entities—way of being in the world.

Sutherland resists the thought that the term 'God' can be translated without remainder into any other terms, even the terms of an objective but secular ethics. The meaning of religious language, he says, is internal to the practices of worship, prayer, reverence, love and humility, which define religious life at its best. The temptation at this point may be to say that these are just subjective attitudes which one may care to adopt. But attitudes are specified by their objects. They point to features of reality to which the adoption of specific attitudes is appropriate. Those features, in the case of an ethical life, are possibilities rather than actualities, and we are not to just contemplate them as pure ideas in a quasi-(or pseudo-) Platonic sense. Sutherland says, 'possibilities define ontologies'. That is, the structures of reality allow the possibility of commitment to transcendent goodness, a commitment which is 'inward' and, as Kierkegaard put it, 'incognito'. I think that in fact for Sutherland they do more than allow; they demand, they have normative significance.

For Sutherland the idea of God is something like, though not adequately translatable without remainder into, the idea of a transcendent order of values, eternal because they do not change with time, and infinite because they are not particular existent objects within the world. They are not entities for contemplation, or entities beyond the world which can enter into causal relations with world-historical events. Therefore, Sutherland is not interested in miracles as physically discernible extraordinary events, in virgin births or physical resurrections. He is interested in identifying features of the world which make inward goodness possible, which demand such goodness, and in the light of which all human motives and goals must be judged.

For such a view, there can be no theodicy, for there is no supernatural person to blame for the ills of the world. But there are objective possibilities and objective demands, which are not just invented by human minds. Alyosha's answer to Ivan is that there is a demand to be compassionate and to love the world, and that it is possible to live so as never to be overcome by evil, even though one might be oppressed by suffering. Bitterness and rebellion are never appropriate. There is no grand coherent metaphysical picture of human life in the world that explains why suffering and evil exist. Here Dostoyevsky, Hume, Kierkegaard, and Camus—about whom Sutherland writes so sensitively and tellingly in *Faith and Ambiguity*—are right to expose the radical ambiguity of human existence.

I believe that a return to Spinoza's idea of necessity might be helpful at this point. It would suggest that there is no positive reason why evil should exist except that the possibilities of evil exist necessarily in God, and they, or some unknown range of them, are necessarily actualised in this or in some or perhaps in all possible worlds. This is not because they are necessary means to good, or because they are freely chosen by a supernatural person, but just because they must be. We cannot 'see why' suffering exists, because there is nothing to see. Ivan cannot 'return his ticket', because he is simply compelled to travel. We cannot see why or how some things might exist by necessity, and in that way we are unable to attain a finally compelling metaphysical picture. But we may at least see that this is a possibility that the structure of reality might allow. The question is how one will react to this situation; that is our 'inward choice'.

In this ambiguous and imperfect world, Sutherland writes that 'the inheritance of theism includes the cultivation of an awareness of the eternal in human life' (*God, Jesus, and Belief*, p. 208). Among the necessities of being, there is a transcendent order of eternal values. Awareness of them takes the form of 'a demand from without', and its main elements are concern for others and humility (importantly, in the way one holds one's beliefs as well as in the way one comports oneself in social life). Human lives are an 'intersection of the eternal and the temporal, the finite and the infinite'. Though Sutherland is perhaps too metaphysically sceptical to follow Spinoza here, one could say (Hegel did say) that the temporal and finite is necessarily and essentially imperfect, even though it essentially expresses part of what God (*qua* impersonal source of beings) is. The eternal and infinite remain in the essence of God, as timeless and spaceless possibilities of being, and they are known by human beings as possibilities to be realised in human lives.

At this point Sutherland finds value in the Christian claim that 'the transcendent has been manifested in time'. For him this will mean (though I am over-simplifying a little here) that the eternal values of compassion and humility have been manifested in a human life. They are not just possible ideals. They can be, and have been, realised in time. There are, however, two important points he makes about any such manifestation. One is that such a thing cannot be established by historical research. Such research is always doomed to be inconclusive and can only arrive at contestable results. Thus, there can be no question of historical 'proofs' of Jesus' divinity or of the claim that he is a 'manifestation of the eternal'.

The other point he makes is that the goodness of Jesus, like the goodness of any human being, must remain incognito. We can never show that when he died he was not overcome by evil or at least by doubt and despair. Yet if his death on the cross was in faithfulness to his vocation it shows something important about the nature of God—it shows that the triumph of good over evil is possible, and it makes possible a certain form of hope and ultimate optimism. This is not hope for some future good, or a belief that everything will in fact turn out for the best historically. Things have certainly not always turned out well since the death of Jesus. Nor is it a hope for future immortality which might compensate for this life's miseries. The idea that somehow future bliss might compensate for or in some way balance out present suffering is not one that appeals to Sutherland. The hope is rather that goodness cannot be defeated, that true goodness is possible, and that the nature of goodness is not worldly success, but a sort of self-renunciation and commitment to action for the sake of goodness alone.

This is coherent with a picture of God, not as a dictatorial sovereign, but as one who experiences suffering or a self-renouncing attention to goodness. Of course, for

Sutherland such anthropomorphism must itself be renounced. There is no supernatural being who dictates or who suffers. This is a picture of the possibility of goodness and its manifestation in the human world, and that goodness will take the paradoxical form of inward self-renunciation and attentive and compassionate love of the world and of others without possessiveness.

There is a significance in the life of Jesus, but it is not one that can be established by historical research, or that requires acceptance of detailed records in the Gospels of his life and deeds. Rather, that person and that life in fact generated in history a startlingly distinctive view of God, or of the possibility of incarnating transcendent goodness in a human life. This possibility assures the triumph of good over evil, even though it does not guarantee a successful outcome, in worldly terms, of human activities. Jesus' death on the cross was not a success in worldly terms—the Kingdom did not come. But it achieved an ultimate hope and optimism, and the faith that eternal values can be manifested in time, and that they give ultimate significance to a human life, however the history of the world goes on.

Seen in these terms, the life of Jesus is not an eruption of the supernatural into the natural world, replete with miracles and physically inexplicable events. It is the revelation of an authentic possibility and demand that human lives should manifest eternal values, that goodness cannot be defeated, and that that the real significance of human lives will be found in an intersection of the eternal and the temporal.

Though the testimony of history will always remain ambiguous, the idea of such an intersection originated in the life of Jesus and in the perceptions his disciples had of him. This idea took a distinctive form, largely because of the history of Israel, its basic values and its forms of life, and was recorded in different forms in the Gospels, as Jesus' early followers sought to express how it seemed to them that the idea was manifested for them in his life and person. That does not provide or require irrefutable evidence of exactly what happened in history. It requires only that at that point in history a distinctive perception of God arose, a new way of living *sub specie aeternitatis*, and a new possibility of manifesting a transcendent order of values in the ambiguities of the temporal world was discerned.

There are those believers—to be honest, probably most believers—who would wish for a greater place for something like 'a power making for righteousness' in history and in human lives, and who would hope for a more unequivocal triumph of the good either in history or in the world to come. Surely, it may be said, goodness cannot be wholly incognito. There must at least be prima facie evidence that a person has not done evil things or harboured vengeful thoughts. Yet it is true that it is hard to detect the innermost motives of the human heart. Even a person who acts outwardly in a wholly good way may be motivated by prudential self-interest. And a person whose life has been marked by intense suffering, whose personality is warped by genetic and neurological disorders, and whose environment has encouraged and rewarded tendencies to personal greed and suspicion, may have done as much as they possibly could to live a good life. In that sense goodness, as the actualisation of the best moral life that one could logically be held responsible for, is indeed incognito—at least if one takes a rather Kantian view of the importance of the inner lives of human beings.

It might also be asked what exactly the importance is of believing that the transcendent has been manifested in time. It might be held that transcendent ideals may exist, even though they have never been fully manifested. It may be enough that humans should strive to manifest them as fully as possible, though humans are doomed always to fail to some extent. Would that really matter, since the triumph of goodness of which Sutherland speaks is not and never will be associated with the elimination of evil from the world or with the elimination of suffering from human lives. Is Sutherland's 'ultimate optimism', which seems to consist in the claim that goodness can persist even in the face of great evil, a realistic form of optimism?

I think there is a sense in which Sutherland holds the high ground here. He looks for no facile material reward for goodness, which must be sought simply for its own sake. The example of a good life lived out in face of great suffering can be morally inspiring and can encourage one's personal efforts to resist evil. Yet it must be confessed that the Christian story does not end with the crucifixion. Stewart always questioned whether the life of Jesus was really a tragedy, as it has often been said to be, since it continues with the resurrection and with an actually triumphant 'return' of Jesus in glory at the end of history. Perhaps human life really is tragic, as a form of life which demands goodness without reward in a world where suffering and evil will never be finally eliminated.

If the Christian story is one for which eternal values are not just possibilities, but are actually manifested in a reality of supreme value; if this is a value which has some, however indirect, influence on the way things go in the world; and if there is a cosmic purpose that finite sentient beings should consciously share in that value in some future state, then there is a more obvious sense in which ultimate optimism about human life is appropriate. But for Sutherland, such imaginative possibilities raise too many problems to be convincing. There are problems about whether humans can survive the death of their bodies, problems about saying that there is some positive purpose in a universe ruled by the laws of entropy, problems about the coherence of the concept of a being which somehow manifests all possible values at the same time, and problems about the causal relation of such a being to the universe.

There are just too many problems of a metaphysical, and therefore undecidable, nature for the ethical lives and responsibilities of men and women to depend upon there being a satisfactory answer to them. It may be that some revelatory forms of personal or public experience could give assurance that such problems could be resolved. But revelation only appeals to some people, and it is not wholly satisfactory to found basic existential decisions about human life on such disputed data. What is clear is that suffering exists, and that morality demands. That clarity should not be obscured by the diverse and conflicting imaginative speculations of metaphysical philosophers or by appeals to revelation which are also conflicting and always subject to reasonable doubt.

In his Isaiah Berlin Lecture to the British Academy in 2004,[2] Stewart expresses this view by distinguishing 'pilgrims' (those who believe there is one clear goal of life to be attained by all) from 'tourists' (who simply seek new and stimulating experiences and purposes) and from 'nomads' (who accept that they are ineluctably bound by space and time, yet seek to preserve and enhance what they find to be of value in their own unique paths through life). Such findings are provisional and fragmented, and suggest a plurality of perspectives, eschewing all bold claims to full and final truth. His recommendation is that the way of a nomad, affirming value and integrity in human life without imposing on others one path to one clearly conceived goal, is the ethical task best suited to the human situation.

If some form of Christian faith is to be defended, Sutherland's contribution to the philosophy of religion must be taken seriously. His view does not necessarily exclude a more metaphysical account, if one could be given. And he would be the last person to think that his view is the final word on the subject of religious belief. Yet he is arguably right in querying any claim that God is a supernatural person who can intervene in history at will, and in refusing to make the hope for a better future an essential or primary motivation for ethical existence. He is right in his insistence that transcendent values are not objects to be contemplated for their own sake, but function to lay down possibilities for temporal existence. He is right to stress that all our insights into such values are provisional and fragmentary. Furthermore, any morally acceptable view of God must find a way of accepting that there is horrendous suffering which is not in any way a means to a greater good, and which cannot be justified by any amount of future happiness. That almost certainly means that God is not one separate person or individual beyond the universe who is wholly benevolent, and who is one of the things that exist, even if a supernatural one.

He is also right in arguing that the significance of the life of Jesus is that it is the originating basis of a distinctive form of the belief that human lives can become manifestations of eternal values, and that in this way finite and infinite can be united in human existence. This cannot be established by historical research, or by claims that physical miracles have occurred. There is no external sign of the vindication of

---

[2] Lord Sutherland, 'Nomad's progress', *Proceedings of the British Academy*, 131 (2005), 443–63.

goodness, which must remain ultimately inward and incognito. But there is an insight that was discerned by the disciples in and through the person of Jesus, that gives rise to an ultimate optimism that good cannot be defeated by evil. The legacy of theism is to preserve this ultimate optimism and a form of life which makes sense of practising reverence, love and humility, by maintaining a commitment to the ethical in an ambiguous world.

When Sutherland considers Dostoyevsky, David Hume, Kierkegaard, Weil and Camus, he claims that they too find in human consciousness a claim to some sort of objective goodness which does not entail over-ambitious theories about the evils of the world as ultimately justifiable or good. Such theories are not what seeing things *sub specie aeternitatis* provide. What that way of seeing provides is the awareness that we need not be defeated by suffering or by despair at human evil. There is something eternal that we can to some extent and in some way manifest in time, and that no evil can defeat—at least in the inwardness of human lives. Such a faith is not separated by an impassable gulf from the lives of those who reject talk of God, but who have an awareness of the inescapable demands of morality. There is a real and vitally important sort of faith that lies on the borderlands between belief and unbelief. Talk of God is a way of preserving that awareness by placing it within a more general way of seeing the world and the possibilities it contains, and talk of Jesus as 'the Son of God' is a way of seeing that this is a possibility that is truly open to men and women. One can, he claims, avoid metaphysical abstractions, and preserve this view of the possibilities and ideals of human lives, and perhaps that lies at the heart of Christian faith.

To say these things, and to say them with the patience and subtlety, the humour and insight that are characteristic of Sutherland's writings, is to increase one's understanding of the phenomenon of religion, to suggest new and penetrating ways of approaching Christian faith in the modern world, and to deepen one's insight into what it means to exist as a human being.

Note on the author: Revd Professor Keith Ward is Regius Professor Emeritus of Divinity at the University of Oxford, and Professor of the Philosophy of Religion at Roehampton University. He was elected a Fellow of the British Academy in 2001.